SCIENCE NOW!

Ann Fullick

Ian Richardson

David Sang

Martin Stirrup

Contents

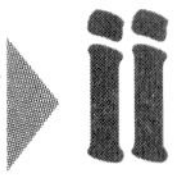

Unit 7 Sound and light

Unit 8 Changing materials

Unit 9 Reproduction

Unit 10 Electricity

Unit 11 Rocks

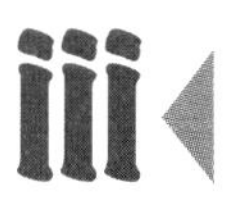

Heinemann Educational Publishers,
a division of Heinemann Publishers (Oxford) Ltd,
Halley Court, Jordan Hill, Oxford, OX2 8EJ

OXFORD LONDON EDINBURGH
MADRID ATHENS BOLOGNA PARIS
MELBOURNE SYDNEY AUCKLAND SINGAPORE TOKYO
IBADAN NAIROBI HARARE GABORONE
PORTSMOUTH NH (USA)

First published 1995

ISBN 0 435 50683 8 (hardback)
99 98 97 96 95
10 9 8 7 6 5 4 3 2

ISBN 0 435 506 82 X (paperback)
99 98 97 96 95
10 9 8 7 6 5 4 3 2

Designed and typeset by Plum Creative

Illustrated by Olivia Brown, Jack Haynes, Lynda Knott, John Lobhan, Dave Marshall, John Plumb, Andrew Tewson, Shirley Tourret, Tony Wilkins and Joanna Williams

Cover design by Miller, Craig and Cocking

Cover photo by Images

Printed and bound in Great Britain by Bath Colour Books, Glasgow

Acknowledgements
The authors and publishers would like to thank the following for permission to use photographs:

p 2: Peter Gould. p 4 T: Barnaby's Picture Library. p 4 B: Peter Gould. p 5: Peter Gould. p 6: Rex Features. p 7 L & M: Peter Gould. p 7 R: J Allan Cash. p 8 T & B: J Allan Cash. p 8 M: Frank Lane Picture Agency/D P Wilson. p 9 T: ANT/NHPA. p 9 BR: Quadrant Picture Library. p 9 BL: J Allan Cash. p 10 & 13: Sporting Pictures. p 14 T: NASA/Science Photo Library. p 14 B (2): Robert Harding Picture Library. p 16 T (4): Heather Angel. p 16 B NHPA/Nigel Sharp. p 18 (2): Peter Gould. p 19 T: Science Photo Library/J Crew. p 19 B: Science Photo Library/Michael Abbey. p 20 L: Heather Angel. p 20 R: Science Photo Library/Dr T E Thompson. p 22 L: Biofotos/Ian Took. p 22 R: Heather Angel. p 23 L: NHPA/J&M Bain. p 23 ML & MR & R: Heather Angel. p 25 L: Frank Lane Picture Agency/M Gore. p 25 ML & MR & R: Heather Angel. p 27 L: Heather Angel. p 27 R: GeoScience Features. p 28 (5): Heather Angel. p 35 T (2): The Natural History Museum. p 35 B (2):GeoScience Features. p 36: J Allan Cash. p 37: Courtesy of Ford Motor Company. p 40: J Allan Cash. p 41: Frank Lane Picture Agency/J Hosking. p 45: David Sang. p 48 T: All Sport. p 48 B (3): Peter Gould. p 50 T: Peter Gould. p 50 A, B, C: Barnaby's Picture Library. p 50 D: Peter Gould. p 51 T: Peter Gould. p 51 M & B: Barnaby's Picture Library. p 52 L: All Sport. p 53: Barnaby's Picture Library. p 56 T: NHPA/ Moira Savonius. p 56 M & B: Heather Angel. p 57 (2): Heather Angel. p 58: GeoScience Features. p 60 TL: Heather Angel. p 60 TR: Richard and Sally Greenhill. p 60 M: Frank Lane Picture Agency/Tony Wharton. p 60 B: Reed Consumer Books Ltd. p 62 TL & TR: Sally and Richard Greenhill. p 62 M: Janine Wiedel. p 62 B: Biophoto Associates/Science Photo Library. p 63 T: Sally and Richard Greenhill. p 63 B: Mary Evans Picture Library. p 64: Reed Consumer Books Ltd. p68: Mirror Syndication International. p 72: Holt Studios/Nigel Cattlin. p 72 inset: GeoScience Features. p 74: J Allan Cash. p 82: Barnaby's Picture Library. p 84: Frank Lane Picture Agency/Silvestris. p 86 T: David Hosking/Frank Lane Picture Agency. p 86 B: Rex Features. p 87 T: Peter Gould. p 87 B: Barnaby's Picture Library. p 88 T: Barnaby's Picture Library. p 88 B: Frank Lane Picture Agency/Mark Newman. p 89: Science Photo Library/Roger Ressmeyer, Starlight. p 90 T: Peter Gould. p 90 BL: Carlos Goldin/Science Photo Library. p 90 BR: Quadrant Picture Library. p 91: NHPA/John Shaw. p 92 L: Mary Evans Picture Library. p 92 R: Barnaby's Picture Library. p 96 T: J Allan Cash. p 96 B: Martin Stirrup. p 97: J Allan Cash. p 98: Mary Evans Picture Library. p 102 T: NASA/Science Photo Library. p 102 B: J Allan Cash. p 103: John Mead/Science Photo Library. p 106: Ann and Patrick Fullick. p 108 T: Heather Angel. p 108 B: Frank Lane Picture Agency/L West. p 109 & 110: Heather Angel. p112 L: Philip Parkhouse. p 112 R: Photri/Barnaby's Picture Library. p 116 (2): Sally and Richard Greenhill. p117: Science Photo Library/Francis Leroy, Biocosmos. p 119 (2) & 120 & 121: Sally and Richard Greenhill. p 123: Science Photo Library/Jim Amos. p 124 (5), 125, 126, 128, 130 (2): Peter Gould. p 131: Science Photo Library/Amy Tristram. p 132, 133, 135(2), 137: Peter Gould. p 138: GeoScience Features. p 139 T (3): The Natural History Museum. p 139 B (2): Peter Gould. p 140 T: Martin Stirrup. p 140 B: GeoScience Features. p 142: Frank Lane Picture Agency/ B Borrell. p 143: NASA/Science Photo Library. p 144: J Allan Cash. p 145: The Geological Museum. p 146 T: J Allan Cash. p 146 B: Barnaby's Picture Library. p 147: Martin Stirrup. p 148: Ancient Art and Architecture. p150 T: GeoScience Features. p 150 BL: Natural History Museum London. p 150 BR: Rida Photo Library/David Bayliss.

The publishers have made every effort to trace the copyright holders, but if they have inadvertently overlooked any, they will be pleased to make the necessary arrangements at the first opportunity.

How to use this book

Welcome to **Science Now**! We have tried to make the book as easy to use and understand as possible. Here are a few notes to help you find your way around.

The book has eleven units. Each one covers one of the big ideas of science in biology, physics or chemistry. Biology units are green, physics units are red, chemistry units are blue.

What's in a unit?

The units are organised into double-page spreads. Each spread has:

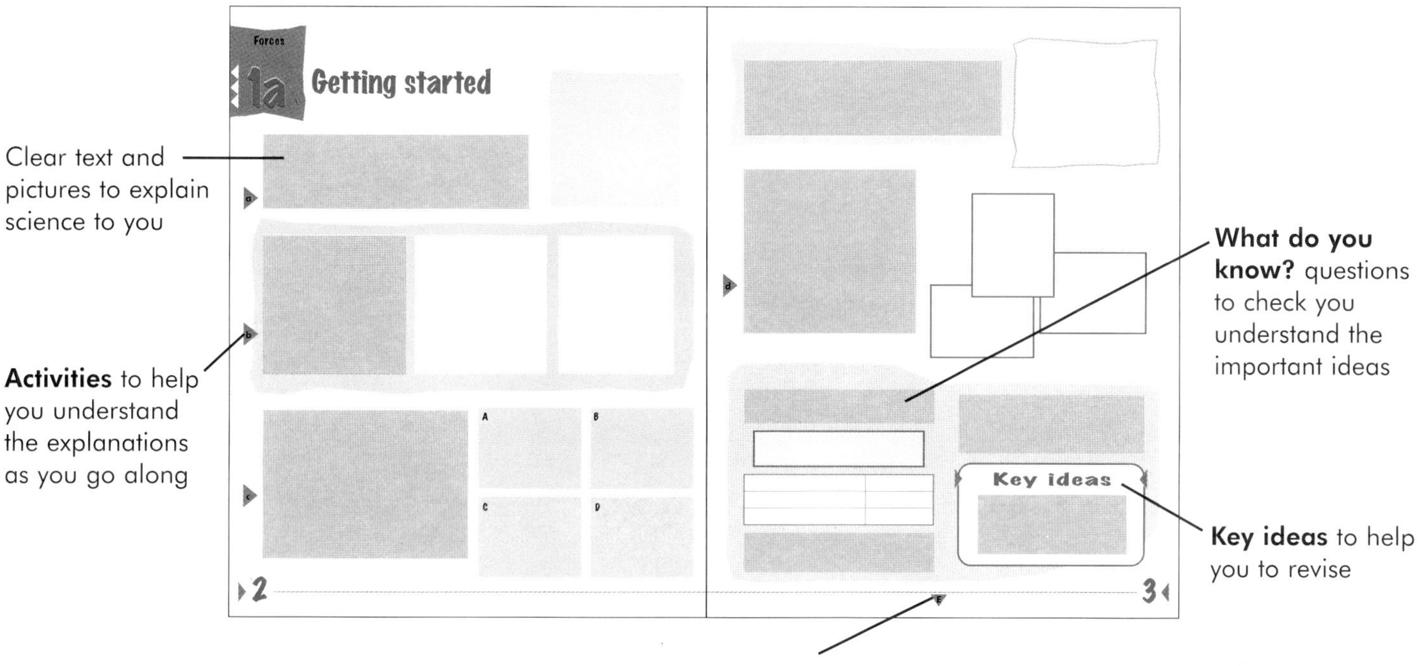

If you finish quickly, some spreads also have a section of interesting new ideas and things to do in the **Extras** pages at the end of the unit

None of the activities needs special equipment or preparation. All the practical activities for the course are in the photocopiable Activities and Assessment Pack which goes with the book.

Glossary

At the back of the book you will find a list of important scientific words and their meanings, so that you can remind yourself quickly of what they mean. If you want any more information you can look at the pages whose numbers are next to the words.

We hope you find **Science Now**! useful in your course. Above all, we hope that you enjoy it!

Getting started

It can be difficult getting out of bed in the morning. You may need someone to give you a **push** or a **pull** to start you moving.

Pushes and pulls are **forces**. When something is stationary, you need a force to start it moving.

a Think of some more examples of pushes and pulls.

The right direction

The shopper provides the push to start the trolley moving. We can show the force by drawing an arrow. The arrow shows us the direction in which the force is pushing or pulling.

b A locomotive provides the pull to start its wagons moving. Draw a diagram with an arrow to show the force on a wagon.

Forces large and small

You need a force to start something moving. Here are some more forces. Each force is making something start moving. Some forces are much bigger than others.

c
1 Put them in order, from smallest to biggest.

2 Say what is being made to move.

3 Say in which direction each force is pushing or pulling.

Balancing forces

In a tug-of-war, the two teams are often evenly matched. They pull on the rope with equal forces, and the rope does not move. The two pulling forces are **balanced**. Eventually, one team gets tired and its force gets smaller. The rope starts to move because the forces are **unbalanced**, and the other team wins. You need an unbalanced force to start something moving.

Measuring forces

We measure forces in **newtons** (N), named after Isaac Newton, an English scientist who lived 300 years ago.

The instrument used to measure forces is called a **newtonmeter**. Some newtonmeters are better for measuring pulls, and others are better for measuring pushes.

Look at the newtonmeters in the photographs.

1 Which newtonmeter is best for measuring a pull?

2 Which is best for measuring a push?

A up to 30 N

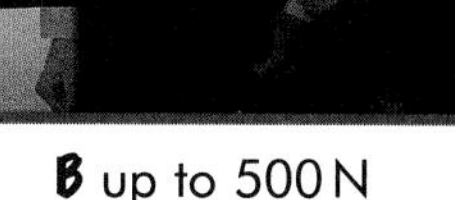

B up to 500 N

C up to 1250 N

What do you know?

1 Copy and complete the following table. Use words from the list below to fill the gaps. You do not have to use all the words.

newtonmeter force (kg) kilograms
dynamometer pull newtons push (N)

two types of force	
units forces are measured in	
used to measure forces	

2 Draw diagrams to show the following forces. Remember to label the arrows.

a the push of a bat on a ball

b the pull of a hand on a drawer.

3 You can easily take a book away from a baby.

a Draw the book, and add arrows to show your pulling force and the baby's pulling force.

b Which of these forces is bigger?

c Are the forces balanced or unbalanced?

Key ideas

You need an **unbalanced force** (a push or a pull) to start something moving.

Forces are measured in **newtons** (N), using a **newtonmeter**.

The pull of gravity

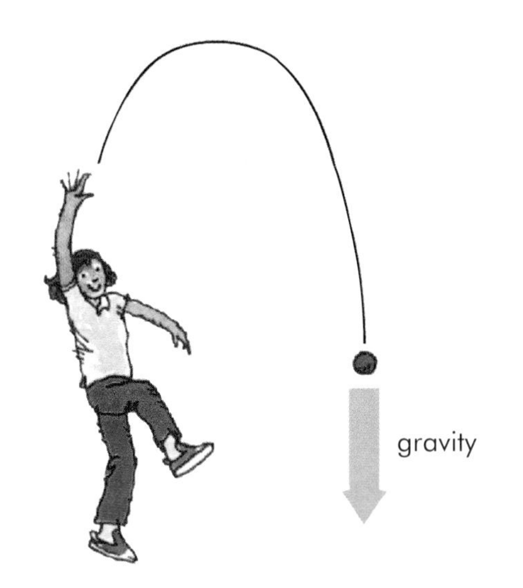

The Earth's **gravity** pulls downwards on everything.

If you throw a stone up into the air, gravity pulls it back down again so that it lands on the ground.

a Use the idea of gravity to explain each of the following. Draw a diagram to show the force of gravity.

1 If you push a rock over a cliff, it falls down to the beach below.

2 If you walk on ice that is too thin, you fall through it.

What's weight?

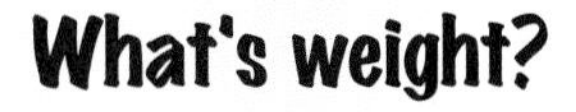

This crane is using a force to lift a heavy steel girder. The pull of the crane must be big enough to lift the **weight** of the steel girder. The weight of the girder is the pull of the Earth's gravity on it. If the chain breaks, the girder's weight will pull it down to the ground.

Because weight is a pull, it is a force. So weight is measured in newtons (N). You can measure weight using a newtonmeter.

b The weight of an apple is about 1 N. Estimate the weight of each of these:

1 two apples **2** ten oranges **3** this book **4** a person.

Space trip

If you go to the Moon, you will notice something surprising. You weigh a lot less. If you drop a stone, it will not fall so quickly on the Moon. If you jump up, you will rise higher before you fall back down.

Some people think that there is no gravity on the Moon, but they are wrong. It is just that gravity there is a lot weaker than on the Earth. To get completely away from gravity, you would have to go far out into empty space, a long way from the Earth and the Moon, and from all the other planets.

What's the matter?

If you go to the Moon, you weigh a lot less. But something about you stays the same. You are still made of just as much 'stuff' as when you were on the Earth. Your **mass** doesn't change.

The mass of anything is measured in **kilograms** (kg). The mass tells us how much **matter** ('stuff') an object is made of. The 2 kg object has twice as much mass as the 1 kg object.

You can pull a trolley along using a stretched rubber band. One trolley has an extra mass on it. You would need a bigger force to make this one keep up.

Which trolley will win the race? Why must both rubber bands be stretched by the same amount to make it a fair race?

Mass and weight

If you weigh an object whose mass is 1 kg, like a bag of sugar, you will find that its weight is 10 N. If you know the mass of something (in kg), you can easily work out its weight (in N). You simply multiply by 10.

In Science, we have to take care to use the words mass and weight correctly. In everyday life, we say a bag of sugar weighs 1 kg. In Science, we should say that the mass of the sugar is 1 kg, but its weight is 10 N.

What do you know?

1 Copy and complete the following sentences. Use the words below to fill the gaps.

force **weight** **gravity** **pull**

The ________ of an object is a ________, caused by the ________ of the Earth's ________ on it.

2 Copy and complete the table. Choose the correct words from each pair.

measured in kilograms	mass/weight
a force	mass/weight
less on the Moon	mass/weight
measured in newtons	mass/weight
stays the same in empty space	mass/weight

3 a What is the weight of a 2 kg bag of potatoes?
b What is the weight of a 50 kg child?

Key ideas

The **weight** of something is the pull of the Earth's gravity on it.

The weight of something is a force measured in newtons (N).

The **mass** of something tells us how much **matter** it is made of.

The mass of something is measured in **kilograms** (kg).

Moving along

These men have solved the problem of how to make the lorries move. They have very strong muscles, and they can pull with a big enough force to start the lorries moving.

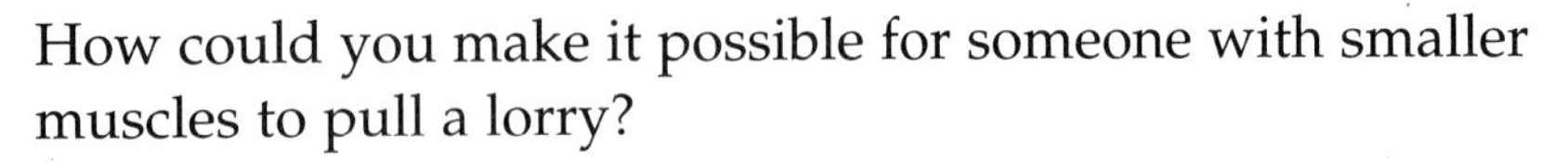

a How could you make it possible for someone with smaller muscles to pull a lorry?

Rubbing along

It is easier to move something if you can get rid of **friction**. You can feel friction when you rub your hands together, or if you slide down the banisters. If something is moving, friction slows it down. Things would move faster without friction.

Friction is a force. We can draw a force arrow to show how friction works. The girl is sliding down the banisters. The force of friction slows her down. The boy is trying to push the rock. Friction pushes back against him so that he cannot move it.

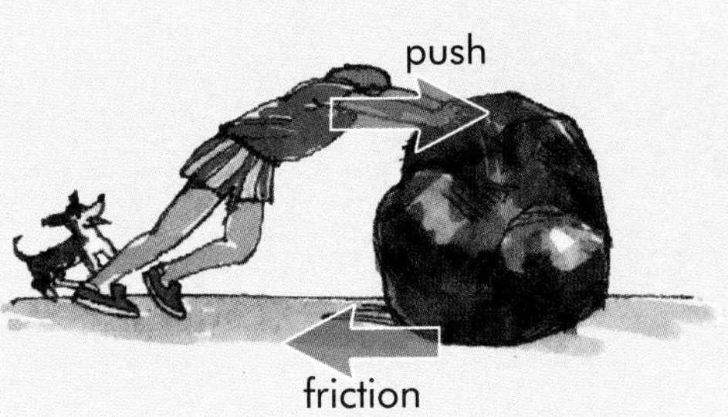

Have you ever tried walking on ice? It is easy to slide on ice, because there is very little friction. If you try to walk, you are likely to fall over. You use the force of friction every time you walk.

Smooth surfaces have less friction than rough surfaces. It is difficult for one rough surface to slide over another.

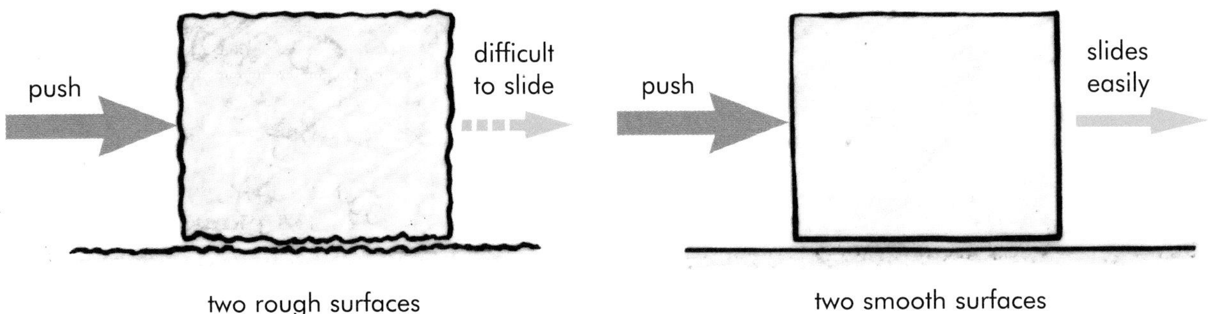

Overcoming friction

Here are some ways of reducing friction, so that it is easier for something to start moving, or to move faster.

How do these things reduce friction?

Roller blades have smooth ball bearings inside the wheels.

In car engines, oil is used to lubricate the surfaces.

A hovercraft floats on a layer of air.

Increasing friction

Friction can be a nuisance, but it can also be very useful. The brakes on a bicycle use friction to help you stop. The rubber pads press on the wheel rim, and the friction slows the wheel down. Car tyres have a pattern of tread to give good grip on the road. If the tyres are bald, there is less friction. The car may slide out of control.

Beating friction

If you want to make something start moving, then you must give it a push or a pull that is big enough to overcome friction. In the diagram, the arrow for the pushing force is longer than the arrow for the force of friction. The push is a bigger force. The forces are unbalanced, so the car starts moving.

What do you know?

1 Copy and complete the following sentences. Use the words below to fill the gaps.

start **surface** **less** **force**

Friction is a ________. It is found when one ________ rubs against another. Friction can make it difficult to ________ something moving. Friction is ________ when surfaces are smooth.

2 Friction can be a problem, but it can also be useful. Study these two pages, and find one example of each.

3 Use ideas about friction to explain the following.

a At a swimming pool, there is water running down the flume (slide).

b It is a good idea to spread grit on an icy footpath.

c Car drivers should slow down on wet roads.

Key ideas

Friction is a force which happens when one surface rubs against another.

Friction can make it difficult to start something moving.

Friction can make it difficult for something to move fast.

Friction can also be useful, because it gives you grip.

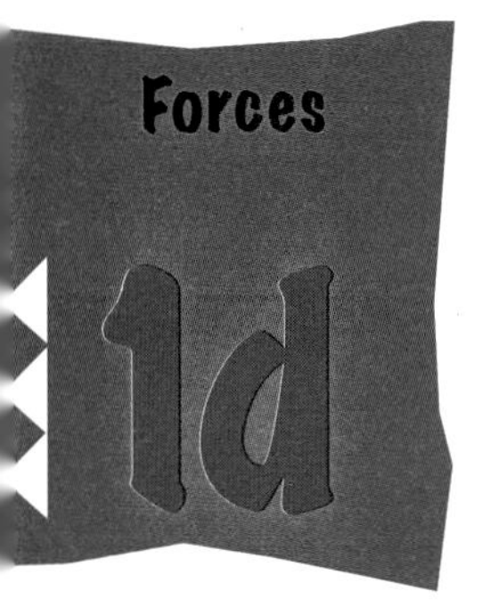

1d Moving in water and air

How fast can you walk? Most young people can walk about 5 kilometres in an hour, which is about 1.5 metres every second. But how fast can you walk in water?

a Describe what happens when you run into the sea, or try to walk in the swimming pool.

Shaped for speed

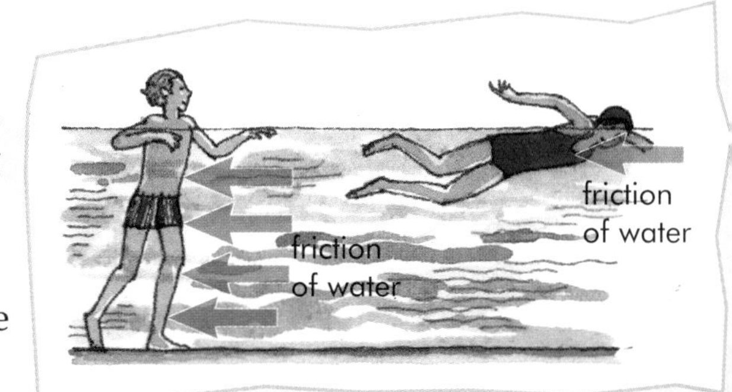

It takes some effort to walk through water because of friction with the water. You can make the friction less by changing your shape. If you swim, it is easier for the water to flow around your body and so you can move more quickly.

This is called a **streamlined** shape. Many fish need to move fast through the water, to catch their prey and to escape from their enemies. They have a very streamlined shape.

b Can you name a streamlined animal that travels fast under water?

Air resistance

Friction with air is a lot less than friction with water, but it is still there. Friction with air is usually called **air resistance**. It slows down all your movements, though you probably don't notice it because you are so used to it.

There is no friction in space, because there is no air. Spacecraft can travel very fast because there is no air resistance at all. The Earth can carry on travelling round the Sun for millions of years without slowing down.

Most big aeroplanes fly high up where the air is thin and the air resistance is less. This helps them to go faster.

Streamlined shapes can make air resistance less, in the same way that they reduce friction in water.

c Look at the picture of the aeroplane. How does its shape help it move through the air?

Using air resistance

A sheet of paper falls slowly to the ground. It falls much more quickly if it is crumpled up into a ball. Air resistance slows down the flat sheet of paper.

Free-fall parachutists know all about air resistance.

If they want to fall faster, they turn themselves head-first so that there is little friction to slow them down.

If they want to go slower, they spread their arms and legs out horizontally, so that there is more friction.

When they are approaching the ground, they open their parachutes. The area of the parachute is very large, so there is a lot of air resistance, and they slow down before they hit the ground.

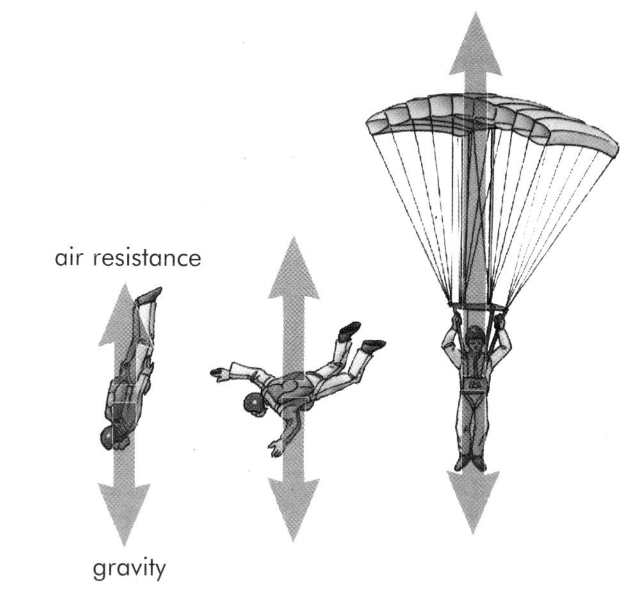

A sugar glider cannot fly. It can only glide. It has flaps of skin between its legs.

How do the flaps of skin help it to glide from tree to tree?

What do you know?

1 Copy and complete the following sentences. Choose the correct word from each pair.
a Air resistance is the force of **gravity/friction** when something moves through air.
b There is **more/less** friction in air than in water.
c Friction makes it **easier/harder** for cars and planes to travel fast.

2 Car designers have to know about air resistance. If they want a car to go fast, they must design it correctly.

A

B

a Which car, A or B, is designed to go faster?
b How does its design help it go fast?

3 Draw a diagram of an open parachute to show how it works. Draw arrows to show the force of air resistance.

Key ideas

There is friction when something moves through air or water.

Friction can be reduced by using a **streamlined** shape.

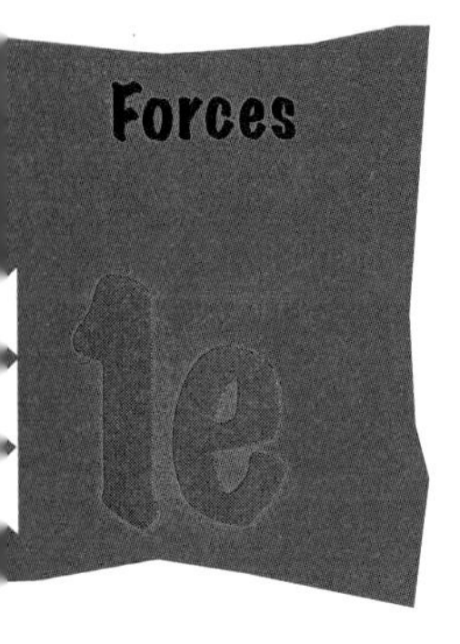

1e With a little help

A champion weightlifter might manage to lift a weight of 2300 N. What's the best you could manage? You might be able to lift someone else in your class, about 500 N. To lift as much as the weightlifter, you need a little help.

a What could you use to help you lift heavier weights?

Levers for lifting

Here is one way to lift a very heavy weight. Use a lever. The child can pull down on the end of the lever. A big force lifts the person in the chair.

A **lever** is a very useful way of lifting something heavy. The longer the lever, the better. The child should push down as far from the **pivot** as possible to be able to lift the heaviest weight.

Getting balanced

A see-saw is a kind of lever. If the two boys weigh the same, they will balance if they sit at opposite ends.

b Where else could the boys sit and still be balanced?

The man doesn't sit at the end to balance the boy. His weight is more, so he must sit nearer the pivot.

c Which way will the see-saw tip if the man moves further away from the pivot?

d Which way should this woman move to balance the see-saw?

Engineers need to understand about balancing forces on levers. The crane in the photo on page 4 has a large block of concrete to balance the weight of the steel girders it is lifting.

More levers

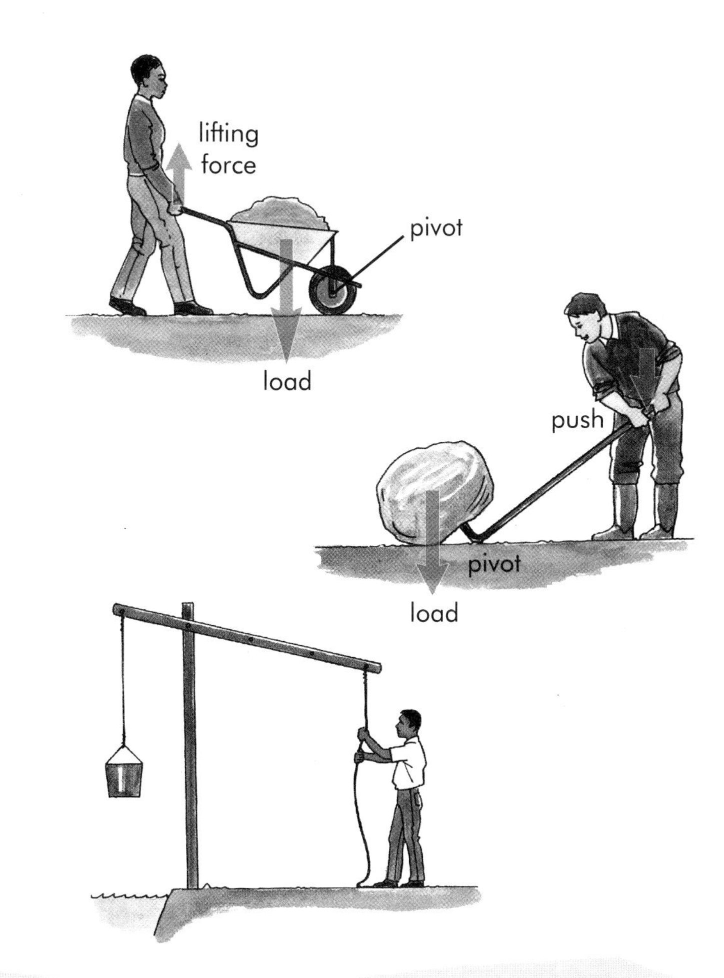

A wheelbarrow is a kind of lever which helps you to lift heavy loads in the garden. The pivot is at the middle of the wheel. Your hands are at the ends of the handles, as far as possible from the pivot.

A crowbar helps you to lift a heavy rock. You use a small force to lift a heavy weight.

The shadoof is a lever which is used to lift water from a river. A small child can hang on the end, and his weight helps to lift a heavy bucket of water.

1 Draw a simple picture of a shadoof, and label the pivot.

2 Draw arrows to show the weight of the water and the child's weight.

3 How could you redesign this shadoof so the same child could lift even more water?

What do you know?

1 Copy and complete the sentences below. Choose the correct word from each pair.

A lever can be useful if you are not **heavy/strong** enough to lift a **heavy/strong** load. You can lift more if you push down near the **pivot/end** of the lever.

2 Jenny has been training in the gym. She finds that the most she can lift is a load labelled 30 kg.

a What is the weight of this load?

b Why can she not lift anything heavier? Use the word force in your answer.

c Explain how Jenny could use a lever to help her lift a 50 kg load.

3 You can use a miniature see-saw to compare weights. A ruler pivoted on a pencil will allow you to compare the weights of two coins.

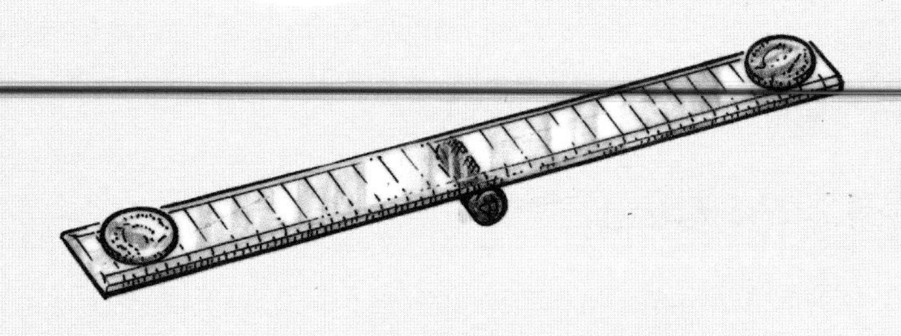

a Suppose you had three coins, and you knew that one was a fake. Its weight is more than the other two, because it is made of different metal. How would you use your see-saw to find the fake?

b If you had seven coins, and one was a fake, how would you find it? You are only allowed two weighings!

Key ideas

A **lever** is useful for lifting heavy objects.

The further a force acts from the **pivot** of the lever, the more it can lift.

EXTRAS

1a The trolley Derby

A

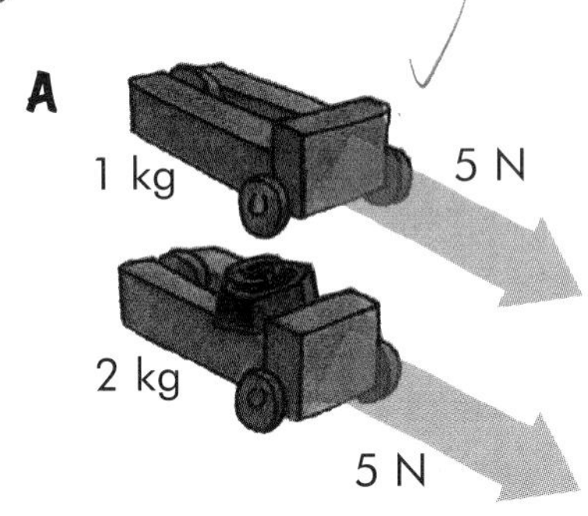

B

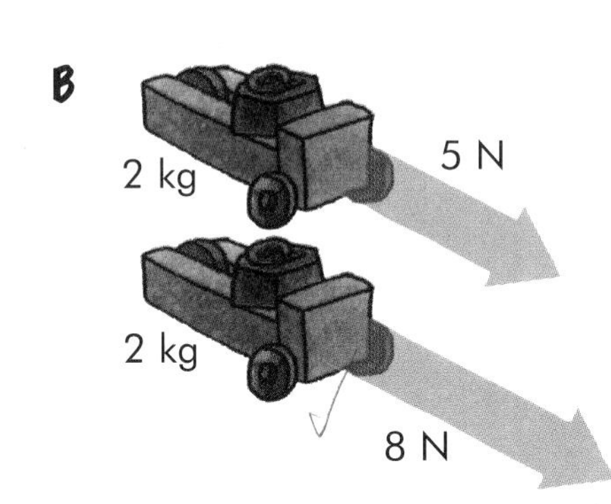

C

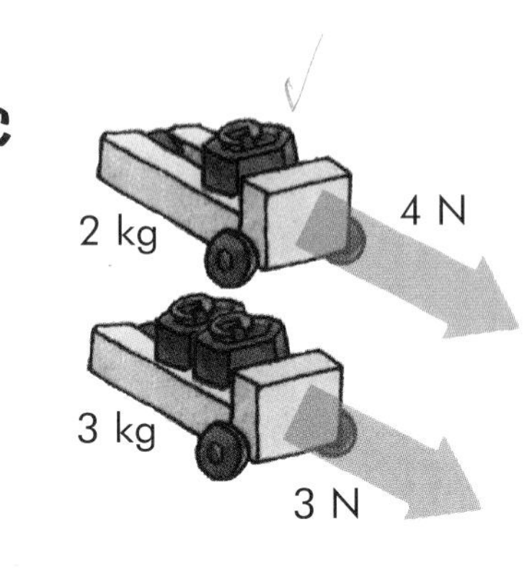

1 The pictures show three trolley races. In each case, say which trolley will reach the end of the bench first. Explain your answers, using the words force and mass.

2 If the three winning trolleys race against each other in the final, which will finish first, second and third? ABC

1b On the Moon

1Kg=10N

When you measure your weight using a newtonmeter, you are measuring the pull of the Earth's gravity on you.

If you go to the Moon, you will find that you weigh much less because the pull of the Moon's gravity is much less. But your mass is just the same as on the Earth. You haven't left behind any of the 'stuff' from which you are made.

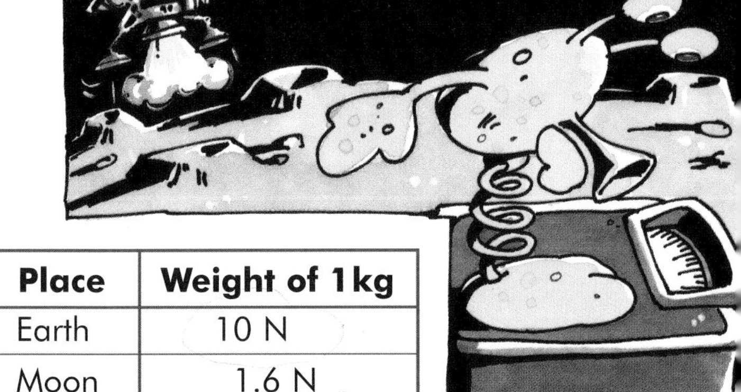

Place	Weight of 1 kg
Earth	10 N
Moon	1.6 N
Mars	3.3 N

1 A space explorer has a mass of 60 kg when he leaves the Earth for a trip to the Moon and Mars. Copy and complete the table on the right, using the table above to help you.

Place	Explorer's mass	Explorer's weight
Earth	60 kg	600N
Moon	60 kg	60N ^ 1/10th
Mars	60 kg	200N ^ 1/3rd

1c Braking

You need an unbalanced force to start something moving. You also need an unbalanced force to stop something moving. The brakes of a car or bicycle provide the unbalanced force needed to bring it to a halt.

1 The diagrams show the forces on a bicycle during a short journey:

a when the bicycle is starting off

b when the bicycle is moving at a steady speed (forces balanced)

c when the cyclist is braking

d when the bicycle is stationary at the end of the journey.

Unfortunately, the diagrams are in the wrong order, and the force arrows have lost their labels. Copy the diagrams in the correct order and add the missing labels.

1d Speedy cyclists

1 Racing cyclists need to understand about friction. Copy and complete the table to explain some of the things they do to make sure that they win.

What racing cyclists do	Why they do it
wear helmets that are pointed at the back	
make sure that their tyres have a lot of tread	
oil the axles of their wheels	
fit new brake blocks before each race	
wear tight clothing and even shave their legs!	

2a Alive or . . . ?

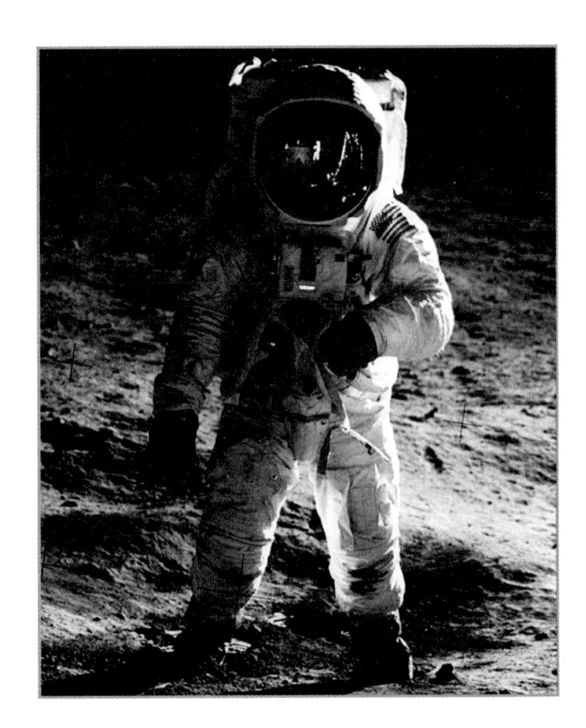

Over 25 years ago Neil Armstrong and Buzz Aldrin stepped onto the Moon. They were very first people to go there. One of their most important jobs was to bring **samples** of the surface back to Earth for testing. Was there life on the Moon?

a Think about how you would tell if a new and strange object was **living** or **non-living**.

Checklist for life

Pictures A and B both show horses, but only the ones in B are alive. How do we know?

b Here is a checklist of things that living things do or need. Copy the checklist and fill it in for the roundabout horses and the real horses.

A

B

	Roundabout horse	Real horse
Does it move some or all of itself?		
Does it need food for energy (fuel)?		
Does it need oxygen?		
Does it produce and get rid of waste?		
Can it grow?		
Can it have **offspring** (babies)?		
Does it feel things and respond to them?		

All living things, big and small, are known as **organisms**. What is true for a horse is true for any **living organism**, including you and me. Living organisms do all the things on your checklist. Non-living things might do some of them, but will never do them all.

Make a checklist like the one on the opposite page and fill it in for:
- an apple tree
- a tiger
- a toy robot
- a fossil.

Which of these things are living and which are non-living?

Meet MRS GREN and her family

Things which all living things need or do are given special names. MRS GREN will help you remember them.

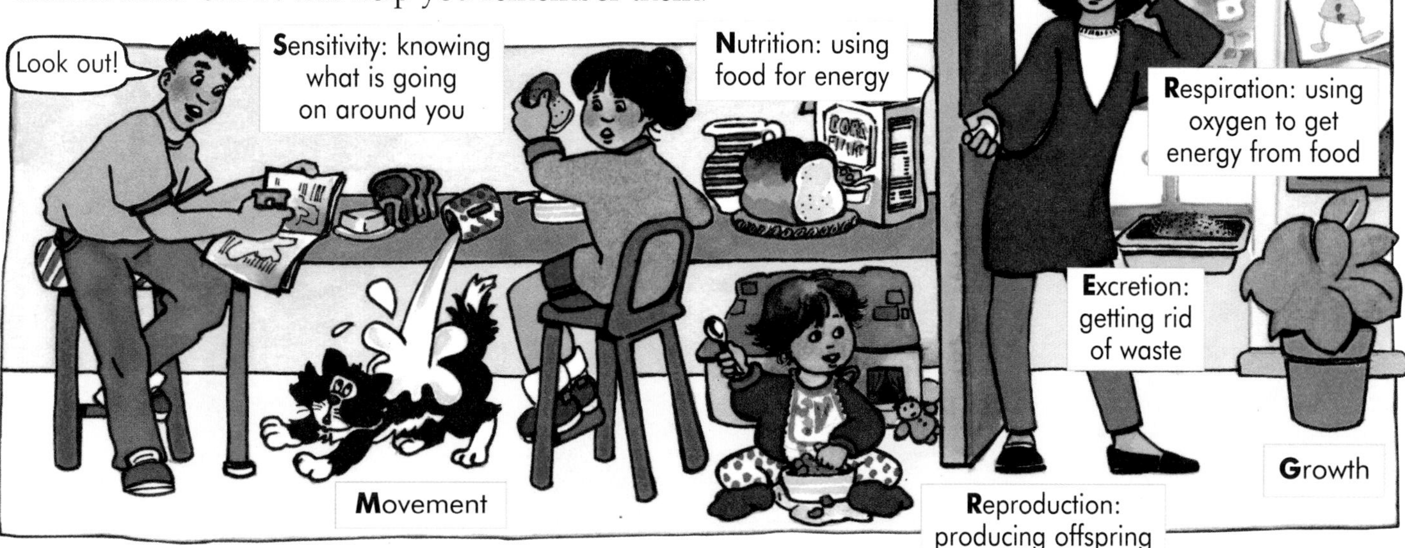

What do you know?

1 Copy and complete the following sentences. Use the words below to fill the gaps.

food **respond** **reproduce** **breathe** **excrete** **move** **grow**

A waxwork is not a living thing because it does not need to ________ air, eat ________ or ________ waste. If you stick a pin in it, it does not ________, and it cannot ________ even its eyes. It does not ________ and it cannot ________.

2 To a visitor from another planet, a car may seem very much alive. List the things about a car which are like a living organism. Why isn't it alive?

3 All the features of life are shown in the picture of MRS GREN. Look carefully and find an example of each one.

Key ideas

Living organisms all carry out the common features of life.

Non-living material may have some of the features of life, but not all.

2b Moving matters

Our senses tell us about the world around us, but it is no good knowing what is going on if we can't do something about it. Doing something usually means moving. We all make thousands of movements every day, for all sorts of reasons.

a Make a list of five different types of movement you have made today. For each one add which parts of your body you moved, and why.

All or nothing?

People move. What about other organisms? All living things move, but they move in lots of different ways. Sometimes only part of an organism moves and responds. Sometimes the movement is so slow or so fast that it cannot be seen by the human eye.

Most plants move very slowly.

Some animals move very fast.

Animal or plant?

We divide the millions of living organisms in the world into two big groups to make them easier to talk about and understand.

- **Animals** are living organisms that move their whole bodies around or move very quickly.
- **Plants** are living organisms that respond to their surroundings by moving only one part of the body very slowly.

Animals and plants move differently, but this is not the only difference between them. There are other important differences.

Who eats to live?

Give a plant sun, air, soil and water and you have a happy, healthy plant. Do the same with an animal and soon it will die. The biggest difference between plants and animals is that animals need to eat, but plants do not. Plants can make their own food using light, air and water.

Which is which?

1 Which of these are animals and which are plants?
2 Make a table to show your answers.
3 Add five more examples of animals and five of plants.

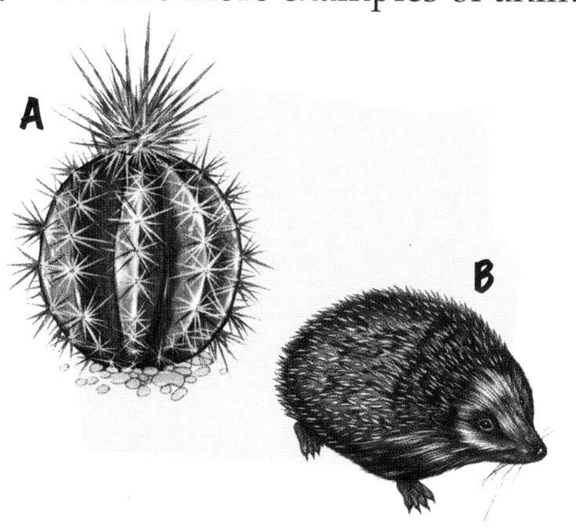

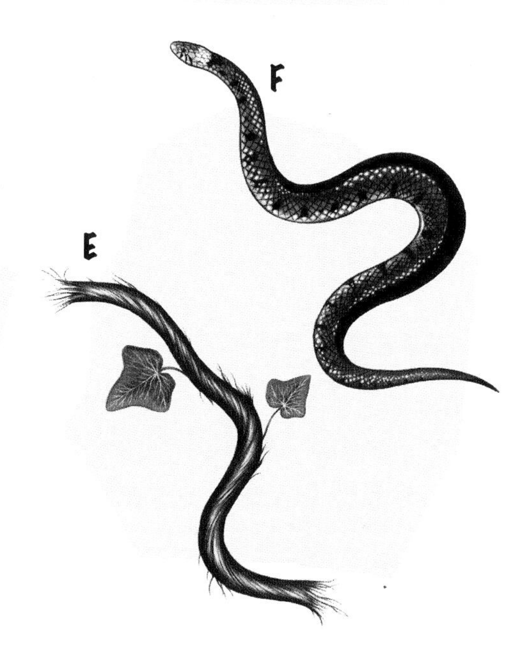

What do you know?

1 Copy and complete the following sentences. Use the words below to fill the gaps.

plants slowly eat bodies
quickly part food

Animals move ________ and they often move their whole ________. They need to ________ to get energy. ________ move only ________ of their bodies and usually move very ________. They can make their own ________.

2 Plants need light to live and they move towards it. They move too slowly for us to see. How could you show that plants really do move?

3 Your little brother runs in with a brightly coloured animal he has found. He is sure it is a flower, but you know it is an animal. Explain how you know whether it is an animal or a plant.

Key ideas

Living organisms can be divided into two large main groups.

These are the **plants** and the **animals**.

2c The Lego of life

Legoland in Windsor Park is full of amazing models like this, containing millions of pieces of Lego. The models range from life-like people to terrifying space monsters, but the bricks used to make them are all very similar. In the same way, all living things are made of building blocks called **cells**. Cells are very small. We need a special tool called a **microscope** to see them.

Animal cells

Each person is made up of several billion cells.

If we look at one of them under a microscope, this is what we see. All animals are made of cells like these.

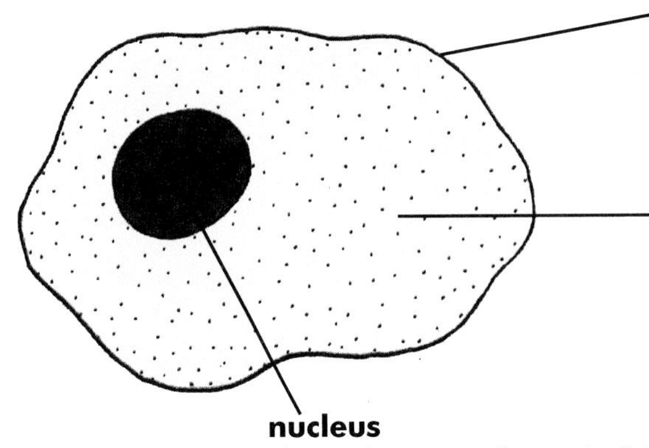

Plant cells

Plant cells are different to animal cells. How can we tell them apart? A quick look at a plant cell under a microscope shows us a big difference.

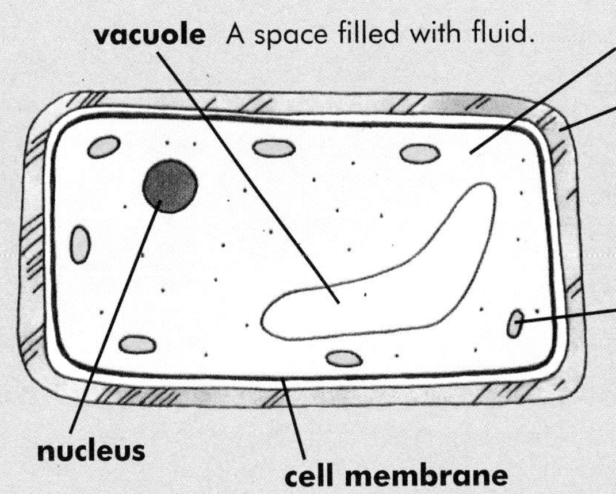

Animal cells are like clear bags of jelly with a blob for the nucleus. Plant cells are like clear bags of jelly in a shoe box, with a blob and also green spots! This is an important difference between plants and animals.

A

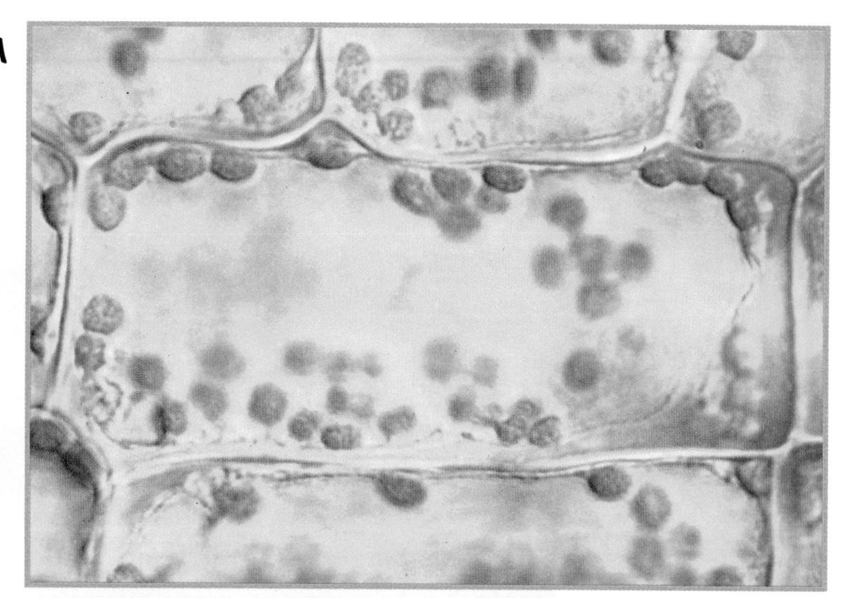

B

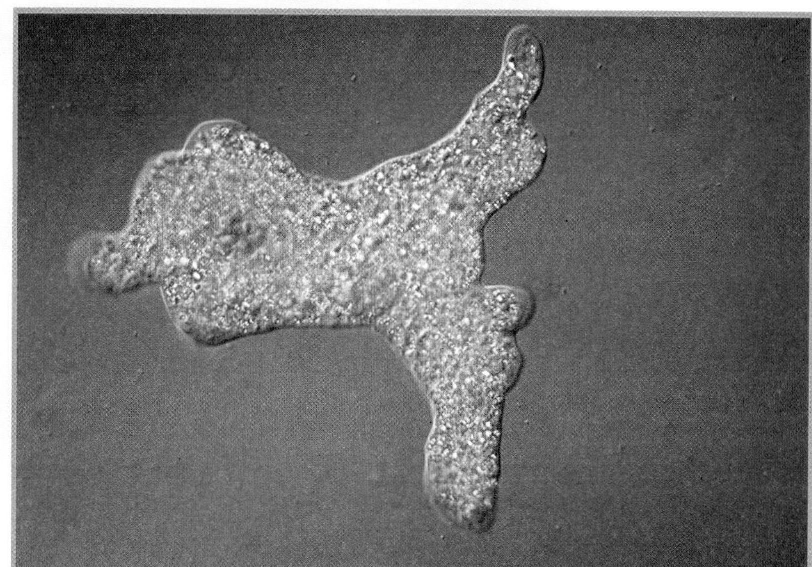

Which is which?

a Here are an animal cell and a plant cell photographed using a very powerful microscope. Decide which cell is which, and then pick out all the main parts in each cell.

What do you know?

1 Draw an animal cell and a plant cell. Label each one carefully.

2 Make a table to show the job done by each part of the cell.

3 Why do you only find chloroplasts in plant cells, not animal cells?

4 Write down the three main differences between plants and animals.

Key ideas

A **cell** is the small single building block of any living thing.

All animal cells have a **membrane**, **nucleus** and **cytoplasm**.

All plant cells have a membrane, nucleus, cytoplasm and **cell wall**. Cells in the green part of the plant have **chloroplasts**.

2d What is an animal?

We all have our own ideas about what an animal is. But when one scientist talks about a particular animal, it is very important that other scientists all over the world know exactly which animal she is talking about.

There are thousands of types of animals in the world. To help us identify them all, we put them into groups. We call this **classifying** them. We start off with big groups, and then split those into smaller and smaller ones.

a Each group is made up of animals that are similar in some way. What do you think are the main groups within the animal world?

Backbones or not?

We look for things that are similar, and things that are different, to help us decide in which group an animal belongs. The two biggest animal groups are animals that have backbones (the **vertebrates**) and animals without backbones (the **invertebrates**).

This animal has a backbone.

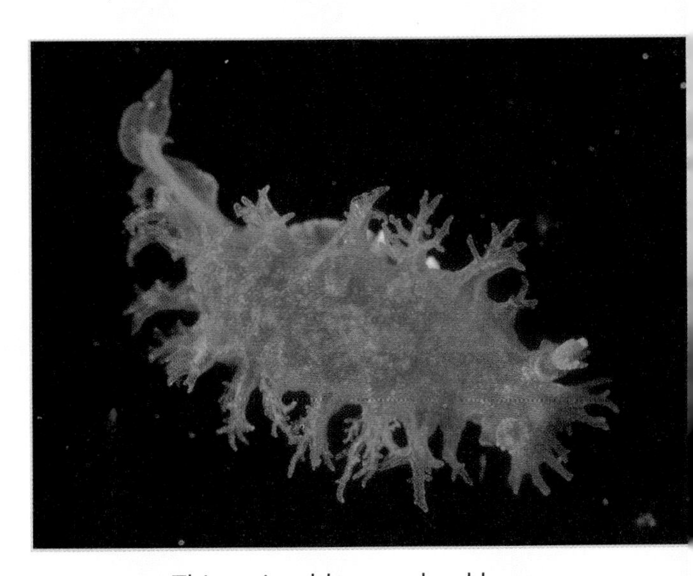

This animal has no backbone.

The vertebrates

Vertebrates have a hard, bony skeleton inside their bodies. This gives them support, protects the delicate organs inside the body and lets them move about.

Vertebrates all have a backbone made up of lots of little bones.

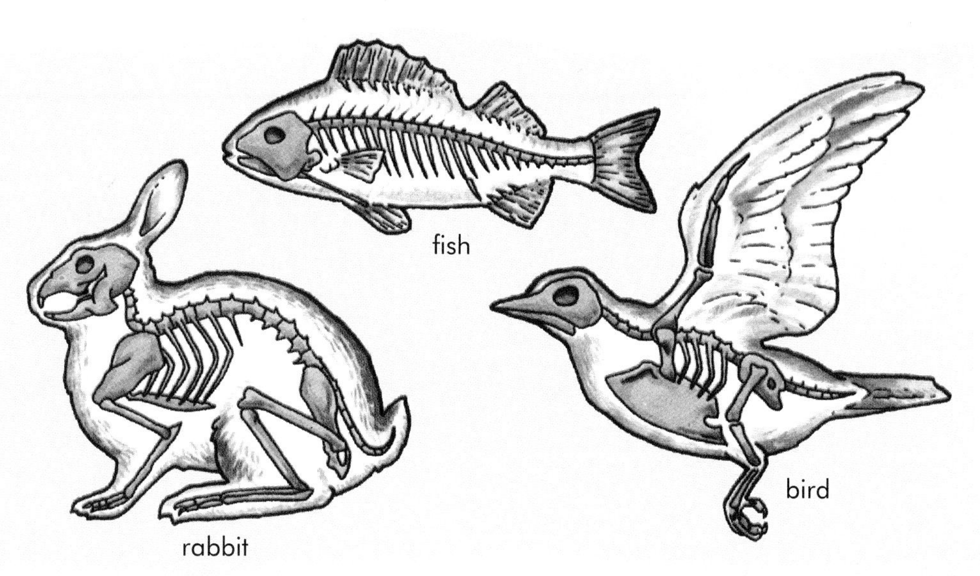

The invertebrates

Invertebrates come in all shapes and sizes. Many of them have soft bodies. Others have shells inside or outside their bodies, or an outer suit of armour. But invertebrates do not have a bony skeleton with a backbone.

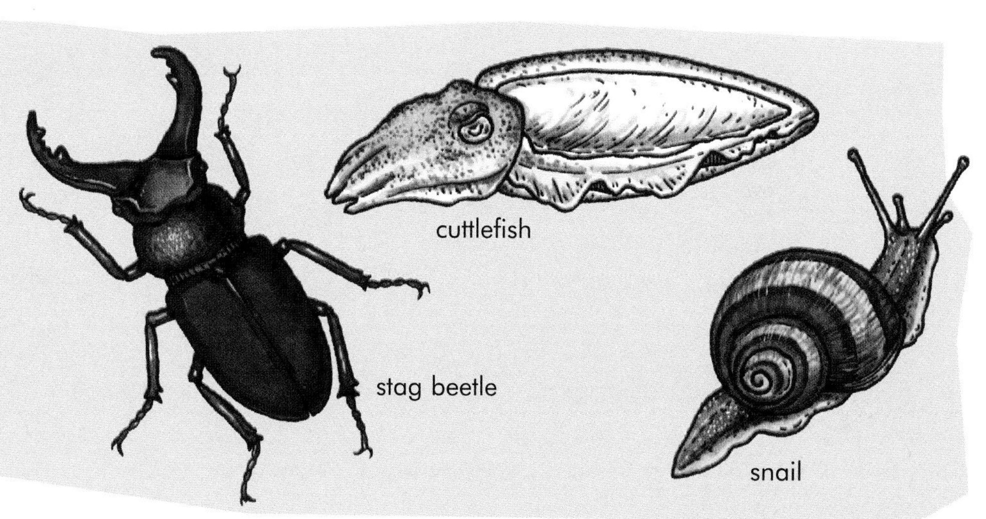

What do you know?

1 Copy and complete the following sentences. Use the words below to fill the gaps.

skeletons snails horses rabbits
backbones soft sparrows without beetles

Vertebrates are animals which all have ________. They have ________ inside their bodies. Examples of vertebrates are ________, ________ and ________.
Invertebrates are animals ________ backbones which often have very ________ bodies. Examples of invertebrates are ________ and ________.

2 Make a list of about 20 different animals. Decide which are vertebrates and which are invertebrates.

Key ideas

We **classify** animals and plants by putting them into groups.

The animal world is divided into two main groups, **vertebrates** and **invertebrates**.

Vertebrates are animals with backbones.

Invertebrates are animals without backbones.

2e Creepy crawlies

There are more animals without backbones (invertebrates) than any other type of animal. They range from simple jelly-like blobs to large and intelligent animals. They can be split up into smaller groups to make it easier to recognise them.

Starfish belong to a group of animals that have a star-shaped body.

Jellyfish have jelly-like bags for bodies, and tentacles to catch their food.

Organising the invertebrates

There are seven main groups of invertebrates. Two of them are found only in the sea. The main examples of these are starfish and jellyfish.

All the other types of invertebrates can be found on land as well as in water.

a Look carefully on the drawing to find them all.

A Flatworms have simple flat bodies.

B Roundworms have thin, smooth, rounded bodies.

C Segmented worms like earthworms have long bodies divided into segments.

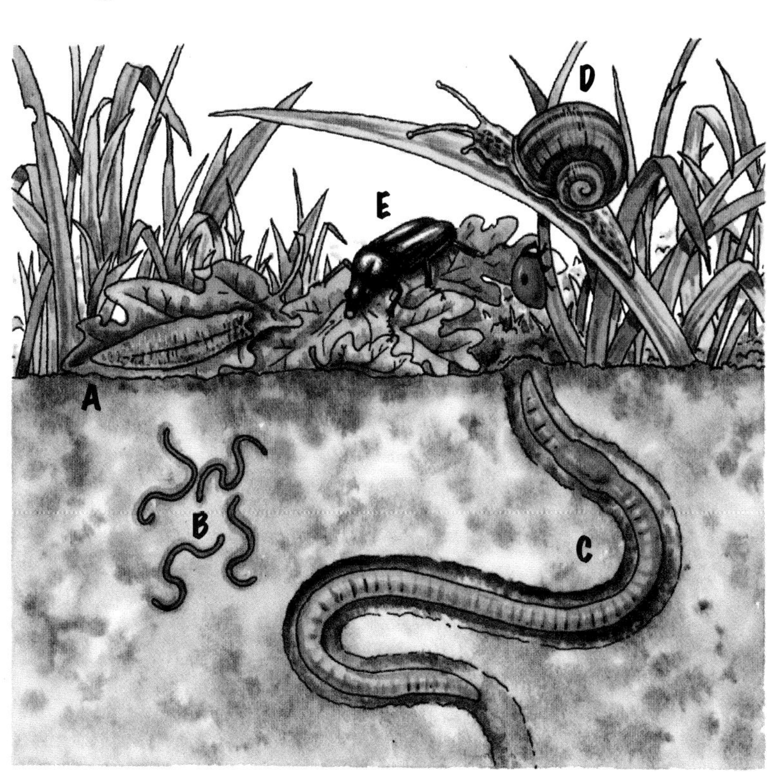

D Molluscs have muscular bodies. Most have a shell inside or outside their bodies. Slugs and snails are molluscs.

E Arthropods have jointed legs and a hard outside skeleton. Beetles are arthropods.

The groups get smaller . . .

Each of these groups of invertebrates can be divided again into smaller ones. There are four main groups of arthropods, as shown in this table:

	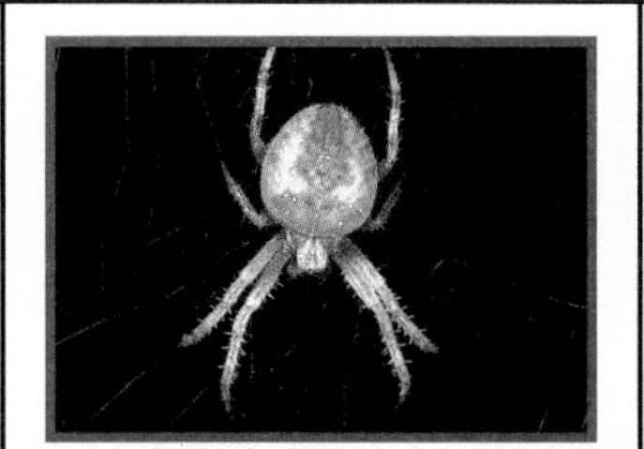		
Crustaceans usually have a hard shell and live in water.	**Spiders** have eight legs and two parts to their bodies.	**Insects** have six legs and three parts to their bodies, and wings.	**Centipedes and millipedes** have lots of segments to their bodies, and many legs.

Sorting them out

1 Copy the table above.

2 Decide which group each of these animals belongs to.

3 Put their letters under the correct heading in the table.

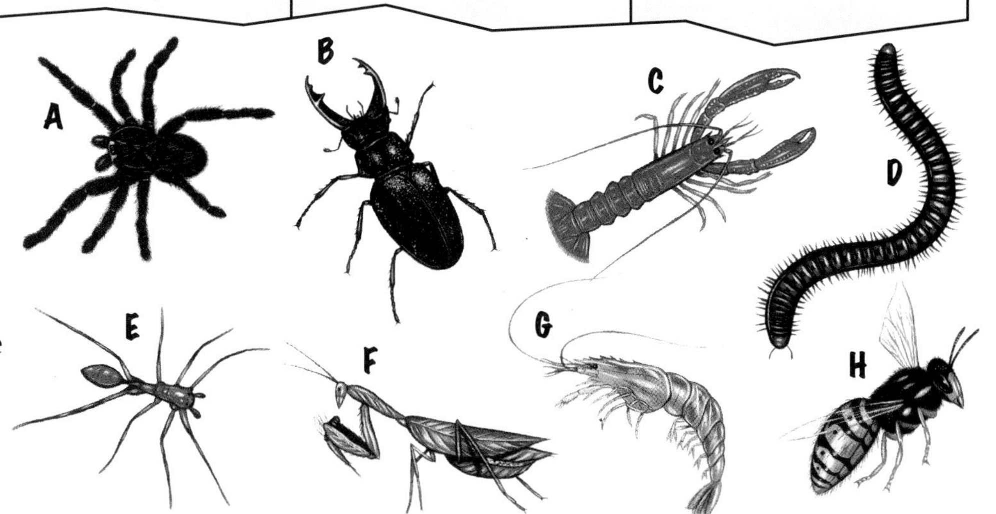

What do you know?

1 Match the names for different types of invertebrates to the right descriptions.

jellyfish	simple flattened bodies
roundworms	muscular body with a shell inside or outside the body
arthropods	long bodies divided into segments
segmented worms	jelly-like bags for bodies with stinging tentacles
molluscs	bodies with five 'arms'
starfish	thin rounded bodies
flatworms	jointed limbs and hard outer skeletons

2 Which of the following are arthropods?

snail **giant octopus** **dung beetle**
tarantula **sea slug** **moth**

Key Ideas

Invertebrates are divided into several large groups: **jellyfish**, **starfish**, **flatworms**, **roundworms**, **segmented worms**, **molluscs** and **arthropods**.

Each large group can be split into smaller ones. For example, arthropods are divided into **spiders**, **crustaceans**, **insects** and **centipedes and millipedes**.

2f Where do we belong?

There are lots more invertebrates than vertebrates, but the vertebrates are much bigger and more noticeable. They can grow larger because they have skeletons inside the body to hold them up.

Organising the vertebrates

Like the invertebrates, we divide the vertebrates into several groups.

A Fish live in water. They have gills for breathing, fins for swimming and scales on their bodies.

B Amphibians like the tree frog have smooth moist skin. They can breathe through lungs and also through their skin. They lay their eggs in water.

C Reptiles like the crocodile have dry scaly skin and they breathe air using lungs. They lay eggs on dry land.

D Birds have feathers and wings and most of them can fly. They have beaks and they lay eggs with hard shells. They often care for their young.

E Mammals have hairy skin and can sweat. The young develop inside the mother's body and are born alive. The mother makes milk in her body to feed them.

The bodies of fish, amphibians and reptiles are at the same temperature as their surroundings. Birds and mammals have their own warm body temperature which stays the same whatever the weather.

The groups get smaller . . .

Each of the main groups of vertebrates can be divided into smaller groups. For example, mammals can be very different. They range from the tiniest shrew to the huge blue whale. Here are some of the main groups of mammals.

 **Whales and dolphins** live in water. They breathe air through a blowhole.	 **Herbivores** eat plants. Cows, sheep, giraffes and horses are herbivores.	 **Carnivores** eat meat and have sharp teeth. Dogs and cats are carnivores.	 **Primates** have forward facing eyes and hands which can use tools.

Sorting them out

1 Copy the table above.

2 Decide which group of mammals these animals belong to.

3 Put their letters into the table under the correct heading.

What do you know?

1 a I have dry scaly skin and lay eggs on land. What am I?

b My babies feed on milk produced by my body. My skin is hairy. What am I?

c I have wings and feathers. My body is always warm and I lay eggs with hard shells. What am I?

d I need water for my eggs. My skin is smooth and moist and I can breathe through it. What am I?

e I need water to live. I have gills, fins and scales. What am I?

2 Human beings are animals. We have backbones so we are vertebrates. Which vertebrate groups do we belong to?

Key ideas

Vertebrates are divided into five main groups: **fish**, **amphibians**, **reptiles**, **birds** and **mammals**.

Each group contains many smaller groups.

2g Pick a plant, any plant

Like the animal world, the plant world can be divided up into smaller groups.

a Make a list of ten different plants. How would you divide them up into groups?

A **Mosses** are small plants that live in damp places. This is because they lose water through their thin leaves, and they can't transport water through the plant. Mosses reproduce by making spores instead of seeds.

B **Ferns** are much bigger plants than mosses. They have strong stems, roots and leaves. Their leaves are waterproof, so don't lose so much water. Ferns have a transport system for water so they don't have to live in damp places. Ferns make spores instead of seeds.

C **Conifers** are usually evergreen. They have thin, needle-like leaves which they keep all through the year. They have a water transport system and waterproof leaves. Conifers produce seeds which are formed inside cones.

D **Flowering plants** reproduce using flowers. The flowers produce seeds inside fruits. Flowering plants have a water transport system and usually have broad waterproof leaves.

Flower power

These four large plant groups can be split into smaller ones.

Most of the plants we see around us are flowering plants. They range from massive oak trees to tiny daisies. Some of them, like trees and shrubs, are woody and can live for many years. Others are what we usually think of as flowers. They only live for a year or two and have bright flowers.

Some flowering plants do not seem to have flowers at all. Grass flowers are green and difficult to spot. But they are still flowers. Grass is a flowering plant.

Horse chestnut trees are flowering plants.

Grass flowers

Using plants

1 Design a garden with a small pond and an area for children to play. Choose plants that will make the garden look nice all year round.

2 Make a plan of your garden showing where you would use mosses, ferns, conifers and different types of flowering plants.

What do you know?

1 Copy and complete the following sentences. Use the words below to fill the gaps.

mosses **cones** **ferns**
flowering plants **evergreen** **seeds**

________ are small plants that produce spores and need a damp place to live. ________ have a water transport system, but also produce spores. Conifers are often ________. They produce seeds which are carried in ________. The ________ ________ produce ________ which grow in fruits.

2 Trees can belong in more than one group of plants. Which are they?

3 Marshes are very wet areas where mosses grow well. Hillsides are drier places where many flowering plants including trees grow, but mosses do not. Explain why mosses grow well in a marsh but not on a hillside.

Key ideas

The plant world is divided into four main groups: **mosses**, **ferns**, **conifers** and **flowering plants**.

The biggest group is the flowering plants. They range from big flowering trees to small grasses.

2h Organise your organisms

A

B

We classify living organisms by looking at the things that are similar between them, and also the things that are different. When we find a strange organism, how can we tell which group it fits into?

Keys

We can find out what an organism is using a **key**. Here is a key you could use on a visit to a zoo to help you identify members of the cat family.

a Choose the animal you want to identify from pictures A–E, and answer 'yes' or 'no' to the questions until you arrive at the name of your animal.

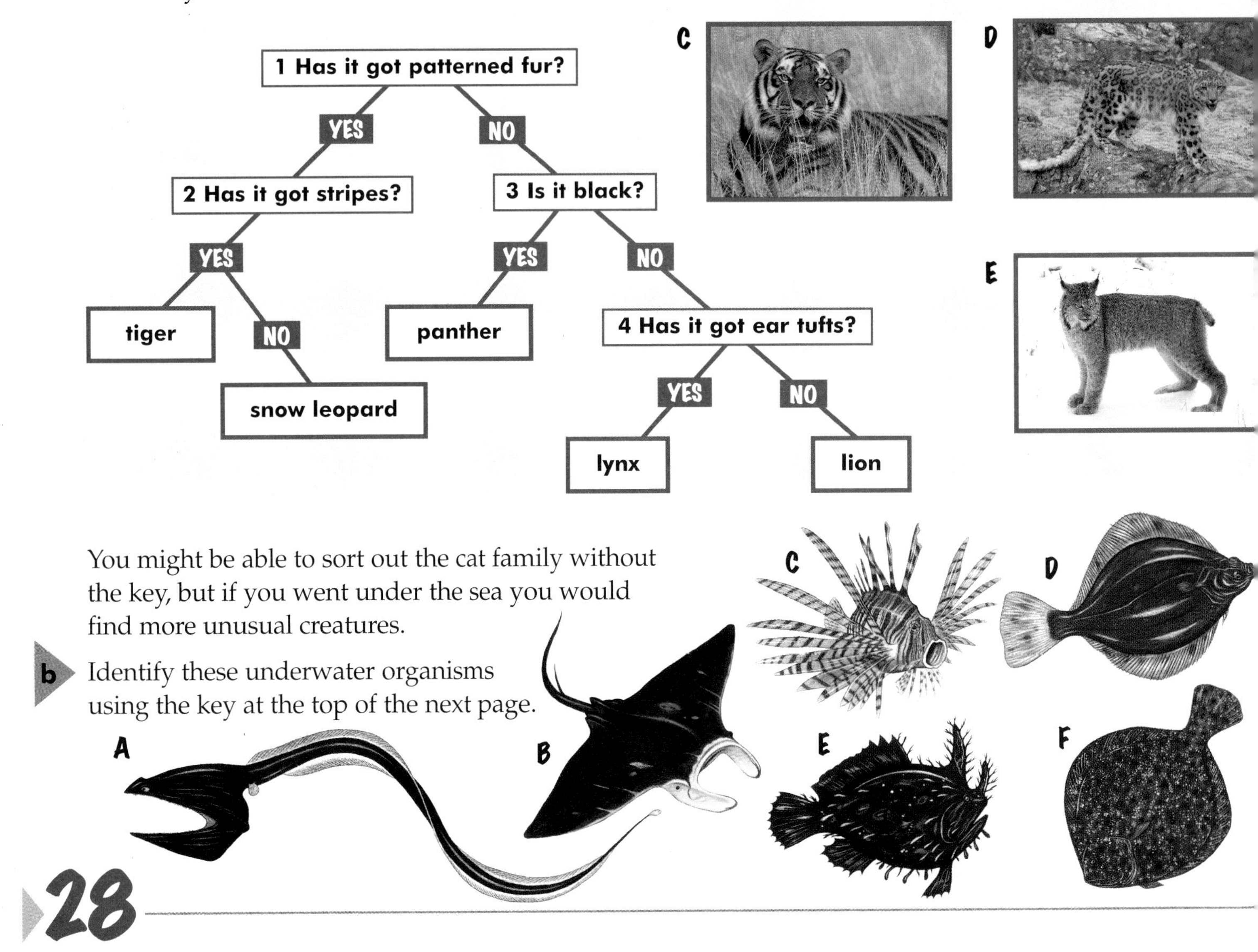

You might be able to sort out the cat family without the key, but if you went under the sea you would find more unusual creatures.

b Identify these underwater organisms using the key at the top of the next page.

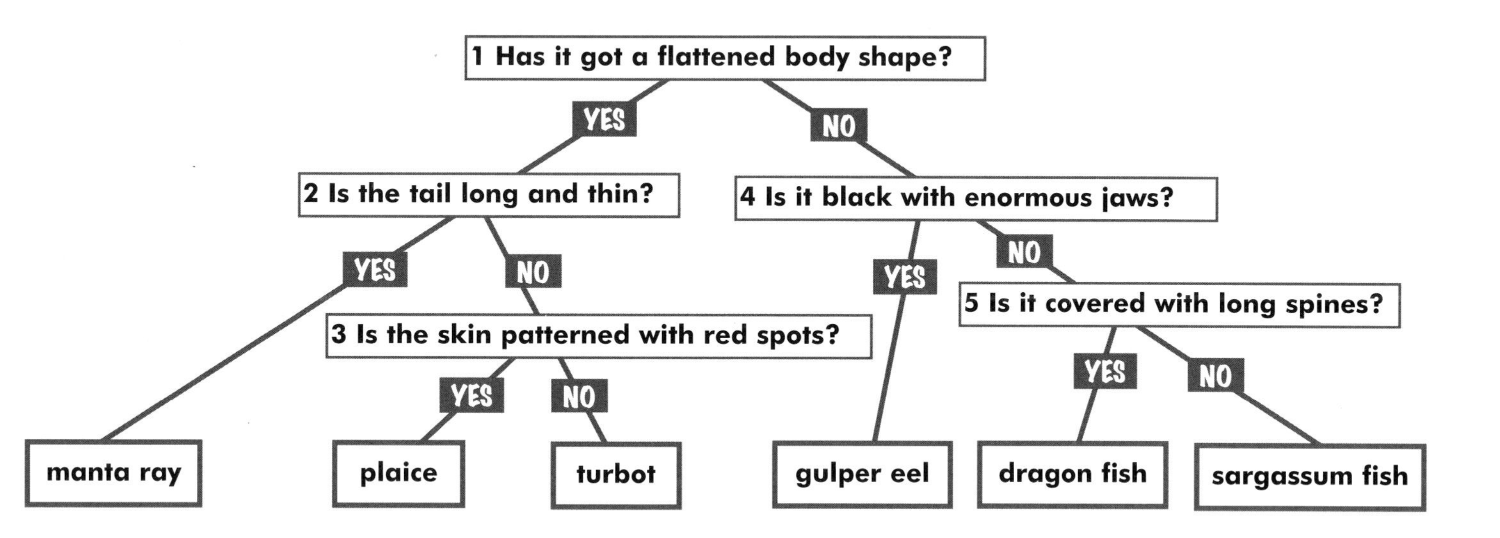

Biologists often use keys when they are working out in the field, observing plants and animals in their natural surroundings. To take up less space, keys are usually printed as a series of questions and answers.

Here is the branching fish key you have just used shown as a table of questions. See if you can still use it.

1 Has it got a flattened body shape?	Yes: go to question **2**	No: go to question **4**
2 Is the tail long and thin?	Yes: manta ray	No: go to question **3**
3 Is the skin patterned with red spots?	Yes: plaice	No: turbot
4 Is it black with enormous jaws?	Yes: gulper eel	No: go to question **5**
5 Is it covered with long spines?	Yes: dragonfish	No: sargassum fish

What do you know?

1 Use the cat key to help you describe the following animals:

a A panther is ________.

b The fur of a tiger has ________.

c A lynx has ________ ________.

d A snow leopard has ________ fur.

2 Take the branching key for the cat family and turn it into a table of questions. Use the fish example to help you.

3 The vertebrates can be divided into five groups: fish, amphibians, reptiles, birds and mammals. See if you can make a branching key that divides the vertebrates up into these five groups. Think about things like the number of legs, whether there are gills or wings and the sort of skin the animal has.

Key ideas

A **key** identifies organisms using simple questions about the differences between them.

2 EXTRAS

2a Living, non-living or dead?

Some objects are obviously living, others are definitely non-living. With some things it is more difficult to tell. For example, wood doesn't do any of the things we expect of a living organism, but it is certainly not non-living because it was once a living tree. Materials like this were once living and are now dead. We are using all or part of their dead bodies.

1 Say which of these things are non-living and which part of organisms:

a wooden spoon **a metal bicycle frame**
a woolly jumper **a pair of rubber wellingtons**
a plastic bowl **a coal fire** **a waterfall**

2 Make a table of the things in question 1 that were once part of living organisms, to show where they came from. For example:

Material	Came from
wood	trees

3 Find other examples of everyday objects that were once part of a living organism, and add them to your table.

2c Planimals?

These little organisms are rather puzzling. They only have one cell, which has to do all the jobs of a living thing.

1 Look at the cells of *Euglena* and *Chlamydomonas* carefully and then copy and fill in this table.

Things found in all cells	Things found in animal cells	Things found in plant cells

Are *Euglena* and *Chlamydomonas* animals or plants? Or should they be put in another group all of their own? Explain your choice.

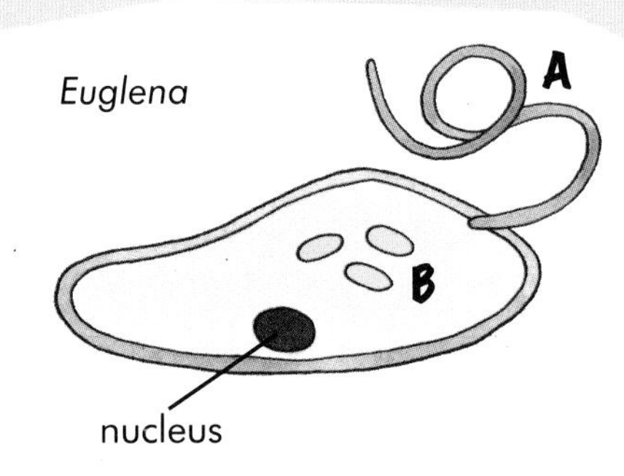

A These flagellae are used to move the whole organism around. It moves quickly for something so small.
B These organisms have chloroplasts so they can make their own food.

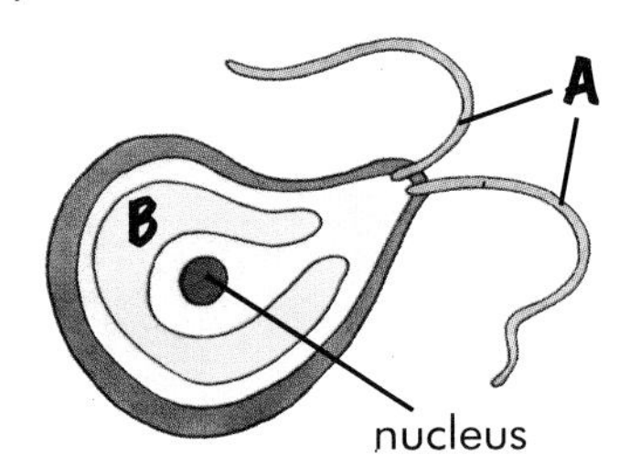

2e Molluscs and more molluscs

The molluscs are a big group of invertebrates which, like the arthropods, can be divided into smaller groups. There are three main groups of molluscs.

octopus

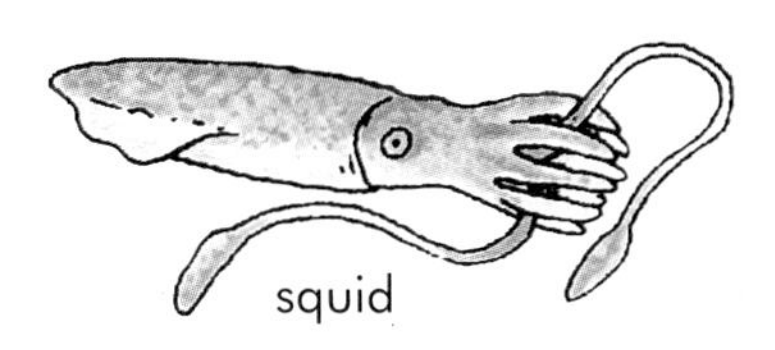
squid

sea slug

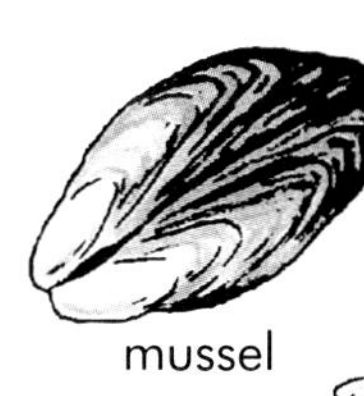
mussel

scallop

snail

whelk

1 Look carefully at these pictures and work out your own way of classifying them.

2 List the animals you think go together, and write down what they have in common. Give each group a suitable name.

Extinct is for ever

The living world contains millions of different types of organisms. All through the history of the Earth, different types of animals and plants have disappeared completely. They have become **extinct**.

This can happen because the weather changes, or because a new disease appears, or because the place where the organism lives is destroyed. Species of animals and plants are becoming extinct more quickly all the time, as human beings have a bigger and bigger effect on the planet.

Macrauchenia

partula snail

seed fern

dodo

Tyrannosaurus rex

1 To which main groups did the extinct organisms in the drawing belong?

2 Make a key for identifying these extinct organisms.

3 What sort of human activities make animals and plants become extinct?

4 Does it matter if animals and plants become extinct? Plan a short talk arguing either that it matters very much, or that it doesn't matter at all.

3a Solids, liquids and gases

You may not have thought much about it, but you probably have a very clear idea about **solids, liquids** and **gases**.

a Think about this picture. Why does it look so odd and dreamlike?

Looking at solids

b **1** Collect some solid objects to look at. Anything will do: a pen, a bottle top, a battery or even your desk.

2 Draw up a table like this. Make a column for each object.

Name of solid object	1 My pen	2 My desk	3
What colour is it?	red	brown	
Is it rough or smooth?	smooth	fairly smooth	
Can you change its shape easily?	no	no	
Can you change its size easily?	no	no	
Is it heavy or light?	light	heavy	

3 Look at your objects and fill in the table.

4 Some of your answers will be different for each object, others will be the same. Write down two things that all solid objects have in common.

What do all solids have in common?

One thing about solid objects is that their sizes do not change. Another way of saying this is that a solid object always takes up the same amount of space.

This space is called its **volume**.

Another important thing about all solids is that you cannot change their shapes easily. These two things are the same for all solids, and are called the **properties** of solids.

How do liquids behave?

Liquids are soft and runny and change shape easily. That's useful when you pour your Coke out of its bottle and into a glass.

Although liquids change their shape easily, their volume always stays the same. Your one litre of Coke will go into all sorts of different containers, but you will never get more (or less) than one litre of liquid.

One litre of Coke will fill...

...four 250 cm^3 glasses

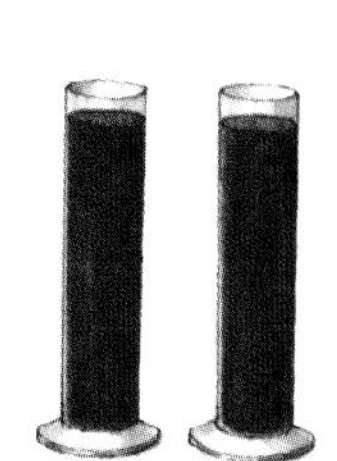

...or two 500 cm^3 measuring cylinders

...or one 1 litre stomach!

1 litre = 1000 cm^3

Gases change shape and volume

Gases change shape too. When you breathe in, the air flows in through your nose to fill your lungs. When you breathe out, this 'lung-shaped' air is pushed out again. Sometimes it becomes balloon-shaped.

If you fill a plastic bottle completely with water, you cannot squash it. Liquids have a fixed volume and cannot easily be squashed.

If you try this with a bottle full of air, you can squash it easily. This is because the volume of a gas does not stay the same. It changes with the volume of the container the gas is in.

you can squash this

you can't squash this

What do you know?

1 Copy and complete the following sentences. Use the words below to fill the gaps.

gas solid liquid shape fixed

Your bones are ________. They are hard and have a ________ shape which supports your body. Your blood is a ________ . It changes its ________ as it flows through your veins. The air you breathe is a ________.

2 A bottle of lemonade has solid, liquid and gas parts. Write sentences like the ones above to describe the three parts and their properties. 'The glass of the bottle is a ________...'

Key ideas

Solids have a fixed **volume** and a fixed shape.

Liquids have a fixed volume but no fixed shape.

Gases have no fixed shape and no fixed volume.

These are called **properties** of solids, liquids and gases.

3b Solid properties

Different objects are made from different sorts of solid. Solids come in many different types, each with different properties. The properties of the solid must be suitable for the object it is being used to make.

a Look at the picture and say why you think these materials have been used for these objects.

Light or heavy?

You need to know about 'heaviness' when choosing the right material for the job.

The frame of a rucksack needs to be made of a **light** material such as aluminium. The less you have to carry, the better. The base of a reading lamp needs to be made of something **heavy** such as iron, to stop it toppling over.

Would you like to carry an iron-framed rucksack?

Weak or strong?

Weak materials will break if you pull or push them. Straw is a weak material, so it is not very suitable for building a house. Bricks, however, are **strong** and are not easy to break.

All building materials such as brick and concrete have to be strong. The foundations at the bottom of a tower block have to be strong enough to support the weight of the whole building without being crushed.

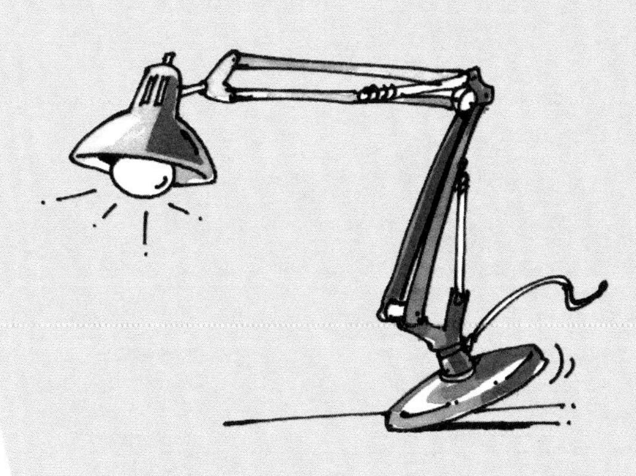

An aluminium base might not be heavy enough for this lamp.

Hard or soft?

A **hard** material will not scratch or dent easily. The hardest material of all is diamond, which will easily scratch glass or metal. Small diamonds are used in rock-cutting drills on oil rigs and in mines.

The 'lead' in your pencil is a **soft** material called graphite. This is so soft that it wears away as you rub it against the paper, leaving a trail of writing.

A scale for hardness

There is a scale of hardness that goes from H1 (the softest) to H10 (the hardest). If you have two objects, the one with the higher number will scratch the one with the lower number.

On this scale, your fingernail is H2.5, a penny is H3.5 and a penknife is H5.5.

Quartz (H7), fluorspar (H4), calcite (H3) and gypsum (H2) are four similar-looking minerals that are often confused. Describe how you could tell these four minerals apart, using just your fingernail, a penny and a penknife.

gypsum

fluorspar

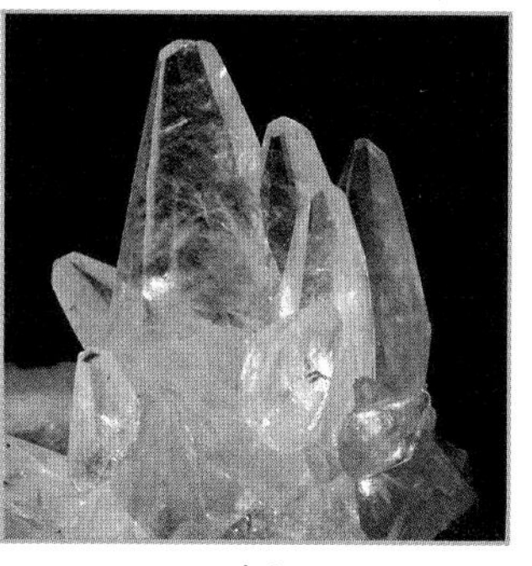

calcite

quartz

What do you know?

1 Copy and complete the following sentences. Use the words below to fill the gaps.

soft **strong** **light**

Bricks and concrete are used for building because they are ________ solids. Aluminium can be used to build aeroplanes as it is very ________ for its size. Pencils use a solid called graphite which is so ________ that it rubs off on paper.

2 Which of the materials in brackets would you use to make the following, and why?

a a ladder (lead, aluminium, rubber)
b a building block for a tall building (expanded polystyrene, brick, lead)
c a rock-cutting drill bit (plastic, glass, diamond)
d an anchor (steel, wood, glass)

Key ideas

Solids can have many different properties. They may be **heavy** or **light**, **weak** or **strong**, **hard** or **soft**.

To choose the right material you have to match these properties against those needed for the job.

3c Useful metals

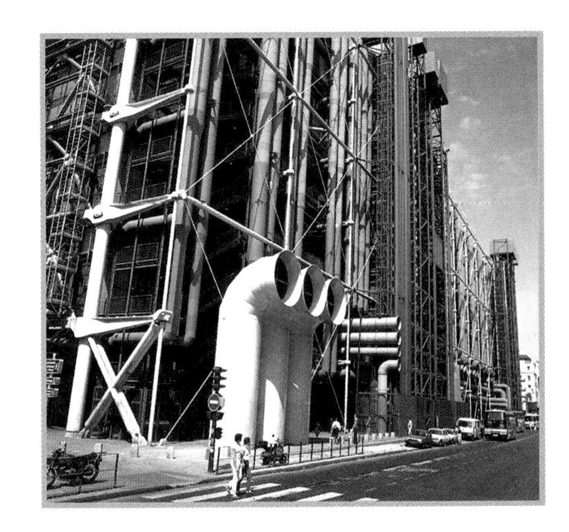

Look around and you will see **metals** in use everywhere.

Sometimes metals are used for their strength. Iron and steel are used to build bridges or support large buildings.

Sometimes metals are used for their hardness. Saw blades made of hardened steel will cut through wood.

a What other useful properties do metals have? Think of some more uses of metals.

Metals conduct electricity

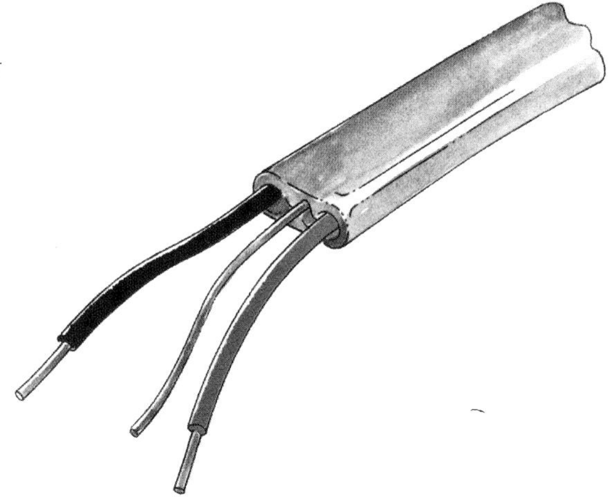

All metals allow electric current to flow through them. They **conduct electricity**. Copper is often used to make wires for carrying electricity because it conducts electricity very well. We say that copper is a very good **electrical conductor**.

You are also a conductor of electricity. If you touched a copper wire that had an electric current flowing through it, you would get a nasty shock. To stop this happening, wires are coated with plastic. **Non-metals** such as plastic do not conduct electricity. They are **electrical insulators**.

Metals conduct heat

Metals also allow heat to flow through them. They **conduct heat**. When you cook stir-fried food in a wok, the heat is conducted to the food from the gas flame through the metal. The metal is a **heat conductor**.

The metal handles also conduct heat, and they may get too hot to touch. You need a heat insulator like padded cloth oven gloves, or you may burn your fingers. Most non-metals are good **heat insulators**.

Metals are shiny

When a metal is new, its surface is shiny. In time, however, most metals become dull. Only a few metals like gold keep their shine and so are used for jewellery. If a piece of jewellery is a cheap imitation, it soon goes dull.

Metals can be shaped

One reason why metals are so useful is that they are very easy to shape. This photograph shows sheet steel being pressed into shape to make car body panels.

Some metals are magnetic

Magnets will pick up pieces of some metals, such as iron or steel. These metals are **magnetic**. Iron (and steel), cobalt and nickel are magnetic.

But not all metals are magnetic. Copper and aluminium are unaffected by magnets. Magnets have no effect on non-metals such as glass or plastic.

Some drinks cans are made from aluminium, and others are made from steel. They often look the same. How could you work out which ones to put in the aluminium recycling bin?

What do you know?

1 Copy and complete the following sentences. Use the words below to fill the gaps.

insulator **conductor** **plastic**

In an electric plug, the pins are made of brass because this metal is a good ________ of electricity. The case is made of ________ as this is an ________ and so stops you getting an electric shock.

2 Make a list of the properties of a typical metal.

3 Graphite is a soft, easily broken, dull black solid that conducts electricity. Look back at your list of properties. Is it a metal or non-metal? Why is graphite unusual?

Key ideas

Metals are useful because they are hard and strong and can be shaped in many ways.

Metals are good **conductors** of both **heat** and **electricity**.

Non-metals are **insulators** of both **heat** and **electricity**.

Metals are shiny.

Some metals such as iron and steel are **magnetic**.

3d 'Heaviness'

Question:

What weighs more, a tonne of lead or a tonne of feathers?

Answer:

They both weigh the same.

So why do people say 'as light as a feather' or 'as heavy as lead'? What is the difference between these materials?

It's all in the volume

The difference between a tonne of feathers and a tonne of lead is that the feathers take up much more room, that is, they have a bigger volume. The lead takes up much less room, so a tonne of lead has a much smaller volume.

If you took the same volume of feathers and lead, the feathers would weigh much less than the lead.

A fair test for 'heaviness'

Snooker balls and ping-pong balls are the same size. Could you tell them apart with your eyes closed? The answer is yes. Snooker balls are much heavier for their size.

To find the real difference in 'heaviness' of different materials, you can weigh pieces of the same size. A 1 cm cube of lead weighs more than four times as much as a 1 cm cube of aluminium, for example.

When you weigh something, you find its mass. The mass of a 1 cm cube of a material is called its **density**. Lead has a high density, but aluminium has a low density.

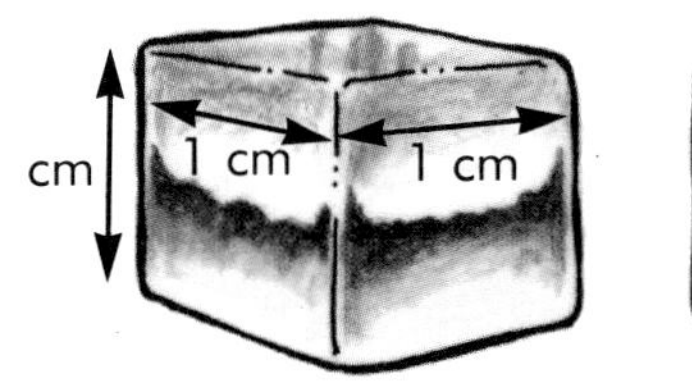

gold 19 grams

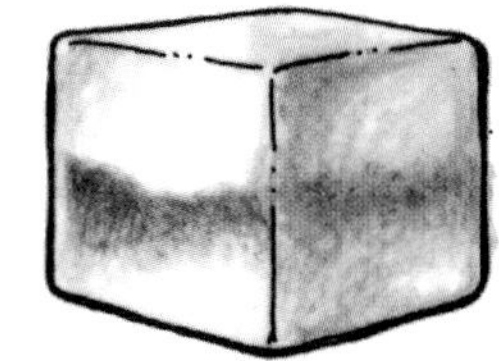
lead 11.6 grams

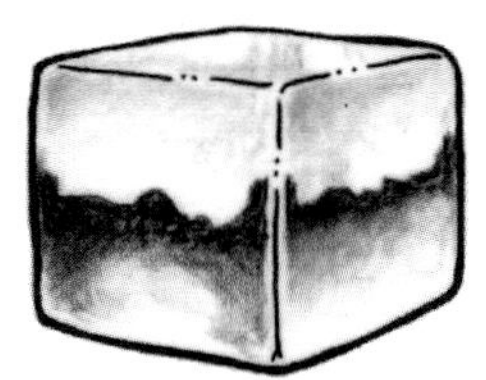
aluminium 2.7 grams

How dense are liquids?

How can you find the density of a liquid? You can't cut a 1 cm cube and weigh it, because liquids spread out.

You can use a measuring cylinder to measure the volume of a liquid in cubic centimetres (cm^3). The density of water is 1 gram per cubic centimetre. Most liquids have densities near to this.

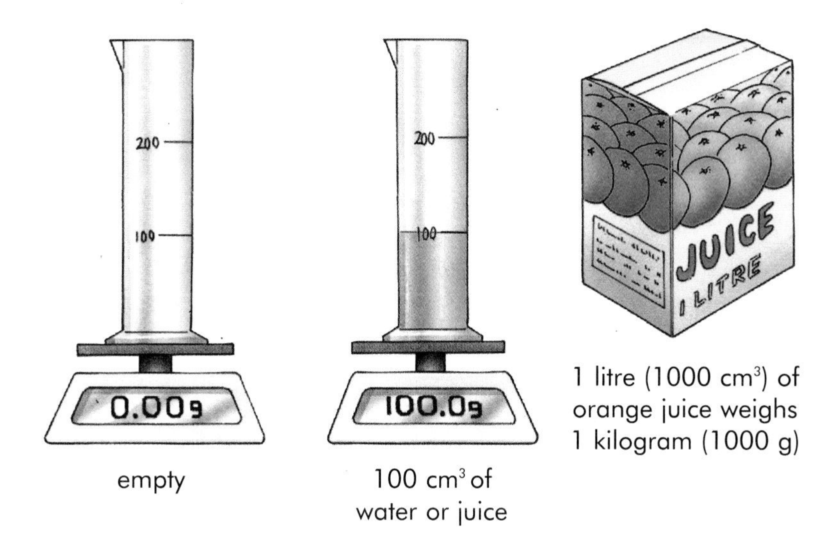

empty

100 cm^3 of water or juice

1 litre (1000 cm^3) of orange juice weighs 1 kilogram (1000 g)

How dense are gases?

A balloon full of air is very light, but a balloon full of water is very heavy. This is because air has a much lower density than water. You would need about 800 cm^3 of air to weigh as much as 1 cm^3 of water. All gases have much lower densities than liquids.

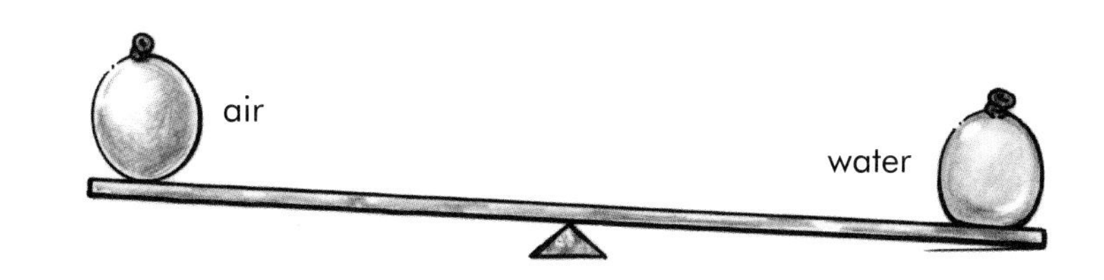

What do you know?

1 Copy and complete the following sentences. Use the words below to fill the gaps.

liquids **volume** **gases** **density**

To find out if one material is heavier than another, you have to weigh pieces of the same ________. The mass of a 1 cm cube is the material's ________. ________ have lower densities than solids and ________ have the lowest densities of all.

2 a Write down all the materials and their densities that have been given on these two pages.
b Rearrange them in order of increasing density.
c Draw a bar chart to show the densities of these materials.

3 Mr Connem the security guard thought he could make a fortune by replacing some gold bars with gold-painted lead bars.
Ms Steindy at the bank raised the alarm as soon as she started to load the bars into the safe.
Explain how Ms Steindy knew that some bars were not gold.

Key ideas

Density is used to compare the 'heaviness' of materials.

The density of a material is the mass of 1 cubic centimetre.

The density of water is 1 gram per cubic centimetre.

3e A model for materials

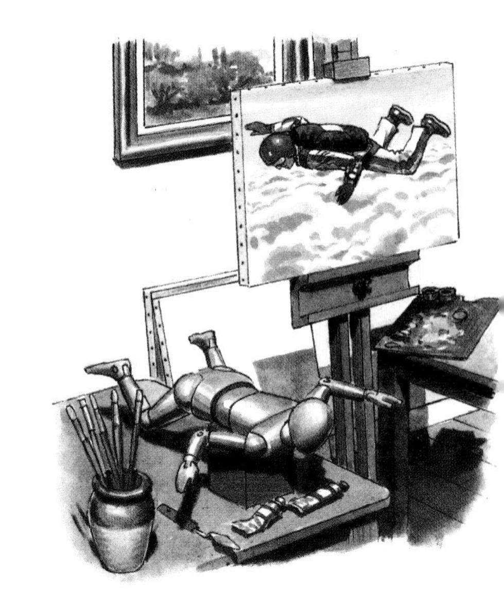

You have seen that materials can have lots of different properties. Why is this?

All materials are made up of billions of tiny particles that we cannot see, even with a microscope. To understand why materials behave as they do, we need to know what is happening to the particles. But if you cannot see the particles, how can you tell what is happening?

The answer is to use a **model** to help you. A model is not the real thing, but it acts like the real thing. Artists often use models like the one here to help them work out how people would look in different positions.

A simple model for materials

Here is a model that helps to explain the differences between solids, liquids and gases.

Sandstone is a hard rock that is sometimes used for building. If you took a block of sandstone and broke it into pieces, you would eventually end up with individual grains of sand.

It helps to think of the grains in sandstone to understand how materials behave. They are a model for the billions of tiny particles that we cannot see.

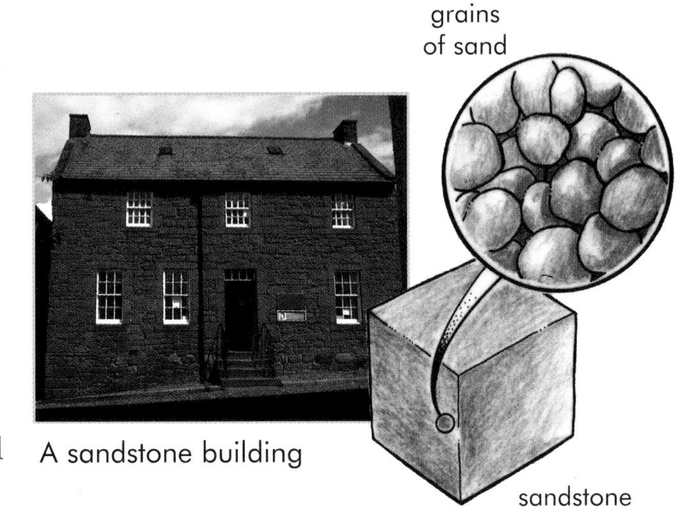

A sandstone building

Explaining the properties of solids

You can't squash a sandstone block, because all the sand grains are packed as close together as possible. The block keeps its shape because the sand grains are all stuck together, so they can't move about easily.

The tiny particles in solids behave in a similar way. The particles must be closely packed, because solids cannot easily be squashed. The particles must also be stuck together, because solids keep their shape.

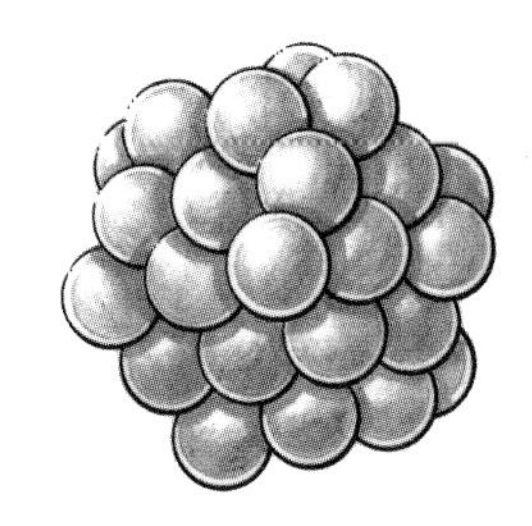

Explaining the properties of liquids

Dry sand behaves in a similar way to a liquid. For example, you can pour it from one bucket to another.

Like sandstone, dry sand is made from closely packed sand grains. But, unlike sandstone, these grains are not stuck together, so they are free to move against one another.

The particles in a liquid must behave in the same way. They must be closely packed, because liquids cannot be squashed. But they cannot all be stuck together, as they must be free to move about, like the grains of sand.

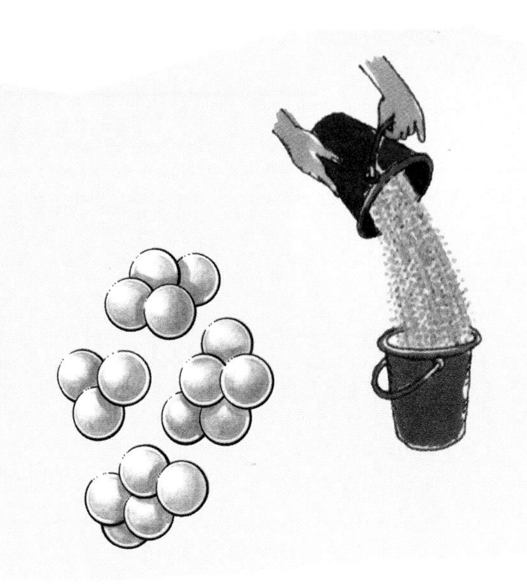

Explaining the properties of gases

Sometimes in the desert, the wind whips up sandstorms. In these storms, individual sand grains are separated out from one another and kept far apart.

In a gas, the particles are also separated out. When you squash a gas, for example air in a plastic bottle, you are pushing the particles close together again.

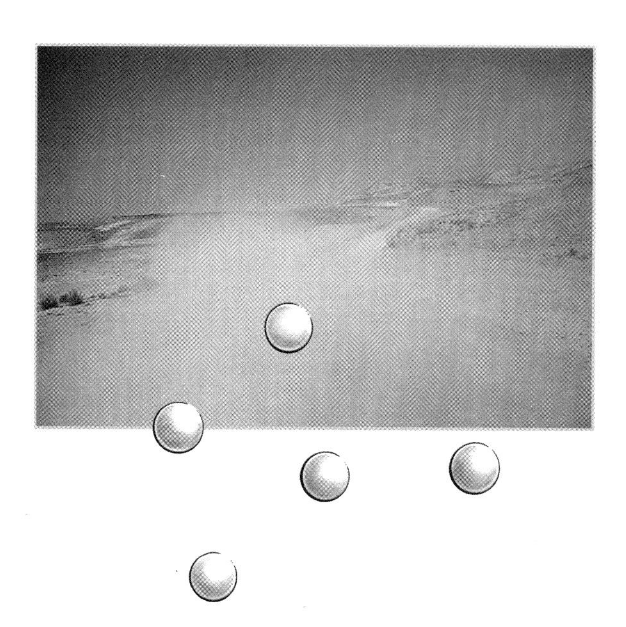

The particle model

Explaining the properties of materials by thinking about how the tiny particles in them behave is called the **particle model**. It is very useful to help us understand how things work.

What do you know?

1 Draw diagrams of the particles in a solid, a liquid and a gas.

2 Use the particle model to explain the following.
a Why can you squash a gas, but not a liquid?
b Why can you pour a liquid, but not a solid?
c Why do solids keep their shape?
d Why are some solids harder than others?
e What happens when you break a solid?

3 You could make a particle model of the pointed end of a crystal by stacking ping-pong balls in a wooden frame.
a Think of a problem with this model.
b How could you improve the model?

Key ideas

Scientists use a simple **particle model** to explain how materials behave.

In solids, the particles are closely packed, and are stuck together.

In liquids, the particles are closely packed but are free to move.

In gases, the particles are far apart.

3 EXTRAS

3a Liquid brakes

The fact that liquids can change shape but cannot easily be squashed is used in the braking system of a car.

1 a Explain how the properties of liquids are used to make these brakes work.

b Sometimes air bubbles get trapped in the oil, and the brakes feel 'spongy' and do not work properly. Use your understanding of gases to explain why this happens.

c The first cars with this sort of brake used water instead of oil. These brakes were not very reliable in the winter.
Can you explain why?

A When you step on the brake pedal, a plunger pushes down on a cylinder of oil.

B The oil is pushed out of the cylinder and along thin tubes which divide and lead to the four wheels.

C At the wheel, the oil pushes into another cylinder, pushing a plunger outwards.

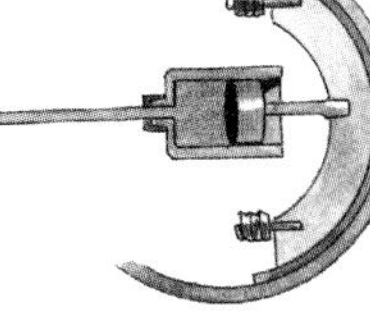

D The plunger pushes the brake blocks onto the wheel.

3c More solid properties

Polythene bottles are often used for oil or household cleaners. Polythene is strong and resists attack from these substances. But if hot oil was poured back into the bottle, it would start to change shape, as polythene is not heat resistant.

Name	Relative strength	Resistance to heat	Resistance to acid	Flexibility	Relative cost
Nylon	high	good	very good	high	medium–hi
Polystyrene	medium	good	excellent	low	medium
Melamine	high	excellent	poor	low	high
Polythene	low	poor	fair	high	low

Here is some information about some common types of plastic.

1 Which would you use for each of the following jobs? Give reasons for your answers.

a a carrier bag **b** a fishing line
c a kitchen worktop **d** a yoghurt pot
e the case of an acid-filled car battery.

2 What additional information would you need before you could decide which material to use for:

a the window of a dolls' house **b** a bottle to contain strong bleach?

3d Changing the weight

A sunshade has a hollow plastic base which comes full of air and so is light and easy to carry. This would not be much help if the wind blew, though. When in position, you can fill the base with water. This makes it much heavier, so it cannot blow over.

full of air:
light and easy to carry

full of water:
now it won't move

1 You can fill your sunshade base up with concrete. Concrete is nearly three times as dense as water.
a How would this be an advantage?
b How would this be a disadvantage?

2 One litre (1000 cm^3) of water weighs 1 kg. An air-filled plastic base weighs just 1 kg. When it was filled with water, it weighed 26 kg. What is the volume of the container?

3e A brief history of comfort

The first bicycles had solid metal wheels. These passed every single bump through to the rider. Then came more comfortable solid rubber tyres. These tyres could change shape a little and even out the bumps on the road surface.

In 1888, John Dunlop invented the pneumatic (air-filled) tyre. The air in the tyre can be squashed quite a lot as the tyre hits a bump, and then it springs back into shape. This gives a far more comfortable ride.

Comfort in the saddle, 1888 style!

1 Use the particle model to explain why air-filled tyres give a more comfortable ride than solid metal tyres.

2 Why would liquid-filled tyres be no better than solid rubber tyres?

Moving particles

Why does a bicycle tyre spring back when you stop squashing it?

If you look at smoke through a microscope, the individual specks are dancing around. Invisible air particles are whizzing around at high speed and crashing into them. Collisions like these give a bicycle tyre its bounce.

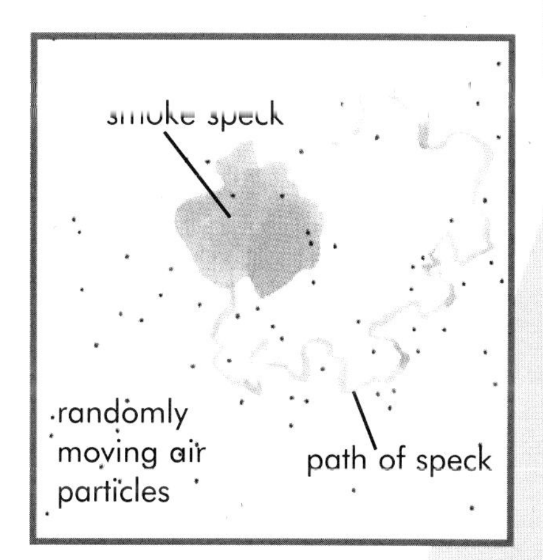

3 Smells are different gas particles mixed in with air. Use this idea of moving particles to explain how smells spread out across a room. (Hint: have you ever been jostled in a crowd?)

4a Getting moving

How fast can you run? You run fast in a race. You might run faster if you were being chased by a bull! You can move even faster on a bicycle, and faster still in a car.

a Look at these pictures of things moving. What is making them move?

Making things move

You need **energy** to make things move. If you hit a tennis ball, you give (or **transfer**) some of your energy to it. You give it **movement energy**.

The girl pedalling the bicycle is transferring energy to the bicycle. The bicycle then has movement energy.

The lazy way

If you pedal your bicycle for a long time, you get tired. You have transferred a lot of your energy to the bicycle. There is an easier way to transfer movement energy to something. You can use a machine to do it for you.

The fan is turning round. It has movement energy. The energy has been transferred to the fan by the electric motor.

b Look at the picture of the motorbike.
1 What is being given energy?
2 How can you tell?
3 What is transferring energy to it?

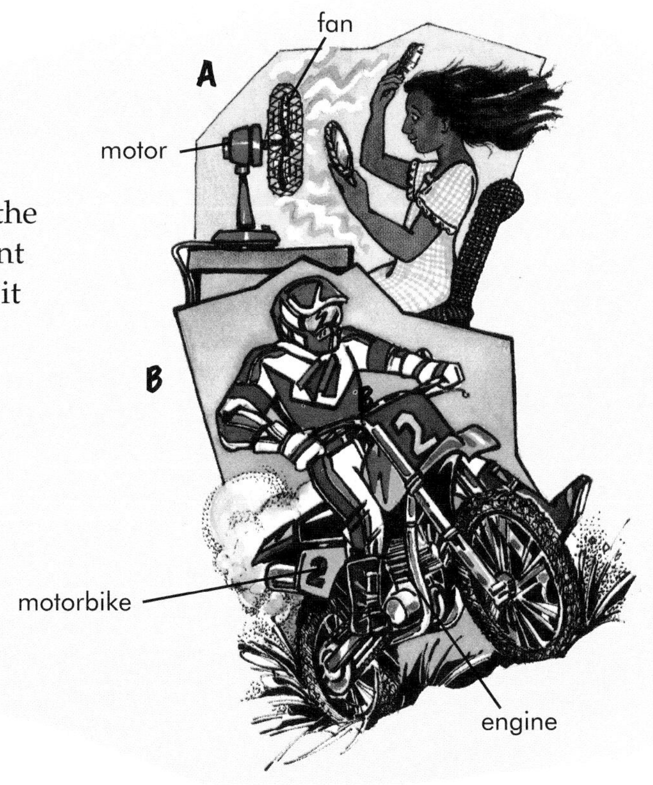

Kinetic energy and kinetic art

These sculptures move around, spraying water everywhere. Water in the pipes makes them move. A sculpture that moves like this is called a **kinetic sculpture**. The word 'kinetic' means 'moving'.

Scientists have a special name for movement energy. They call it **kinetic energy**. Anything that is moving has kinetic energy.

What do you know?

1 This hamster spends a lot of time making its wheel go round. Copy and complete the following sentences. Use the words below to fill the gaps.

movement turning transferring kinetic

The hamster is _______ energy to the wheel by _______ it. The wheel has _______ energy. The scientific name for this kind of energy is _______ energy.

2 Look at the picture of the kinetic sculptures again. Where do they get their kinetic energy from? How could you make them go the other way round?

3 Look back at the things that are moving in the pictures at the top of the opposite page. Which do you think has the most kinetic energy? Which has the least kinetic energy? Give reasons for your answers.

Key Ideas

Everything that is moving has **energy**. We call this type of energy **movement energy** or **kinetic energy**.

To make something move, you have to **transfer** energy to it.

Getting warm

You need energy to make things move. You need energy to do other things, too. You can use your energy to make things hot.

If you rub two blocks of wood together, they get quite warm. If you had the skill of some traditional peoples, you could start a fire like this.

Heat energy is a kind of energy. If the weather is cold, we need to supply heat energy to warm up our homes. We use heat energy to cook our food, and for hot water.

a What sources of heat energy can you think of?

Energy transfers

The cat is enjoying the heat from the fire. Heat energy is being transferred from the fire to the cat. We can draw an **energy transfer diagram** to show this transfer. The arrow shows that heat energy in the fire has been transferred to heat energy in the cat.

b The other two pictures also show transfers of heat energy. Draw energy transfer diagrams to represent each of these.

Heat from fuels

One way we can get heat energy is by burning things. A substance that we burn to release heat energy is called a **fuel**. The pictures show some examples of different fuels.

c Make a list of the different fuels shown in the pictures. Can you think of some other kinds of fuel to add to your list?

Cold coal

A piece of coal is not hot. It does not contain heat energy. It stays cold until we set fire to it. We know that it has energy in it, because we can get the energy out by burning it.

We say that the coal is a **store** of energy. It is a store of **chemical energy**. Most fuels are energy stores. We can include 'chemical energy' in the energy transfer diagram for the cat in front of the fire.

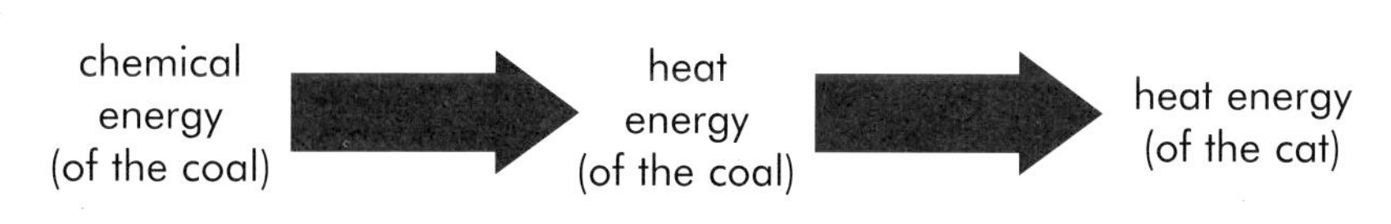

Heat and light

The flames from a coal fire or gas stove give light as well as heat. Some of the stored energy is being turned into **light energy**. Here is an energy transfer diagram that shows this. The arrow splits in two, because we get two types of energy, heat energy and light energy.

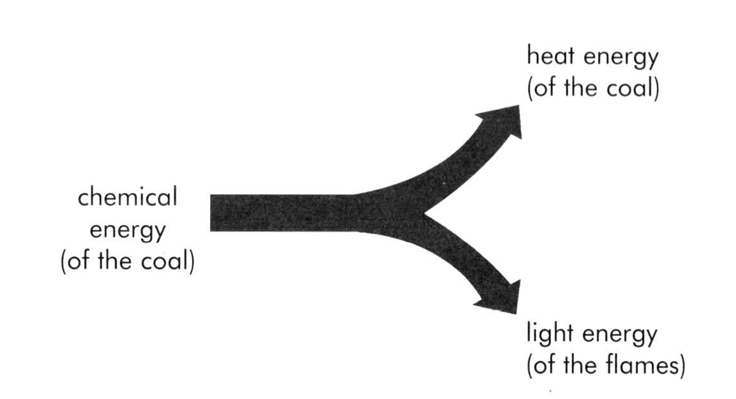

What do you know?

1 a What fuels do you use in your home to give you heat energy?
b Make a table like this to show what you use fuels for.

Room	Fuel	Use
kitchen	gas	cooking

2 Some barbecues use charcoal as the fuel for cooking food. Copy and complete this energy transfer diagram for a barbecue.

______ energy (stored in ______)

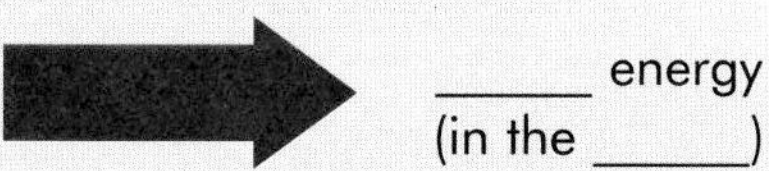

______ energy (in the ______)

3 Copy these sentences that explain what is meant by a fuel. Choose the correct word from each pair.

A fuel is a useful **type/store** of **chemical/heat** energy. The energy is **stored/released** when the fuel is **moved/burnt**.

4 Here are some types of fuel. Can you suggest who might find them useful? How might they use the fuels?

[illegible] [illegible] [illegible]

Key ideas

Something hot has **heat energy**.

A fuel is a store of **chemical energy**. When it is burnt, heat energy is released.

We use **energy transfer diagrams** to show how energy is transferred.

Body fuel

Where do our bodies get the energy they need? They need 'fuel' too.

a What sort of fuel do our bodies use?

Energy in food

Food is the fuel supply for all of our activities. Food is a store of chemical energy. We need energy from our food so that we can move, and so that our bodies keep warm. The chemical energy in our food becomes kinetic energy and heat energy.

INGREDIENTS

MAIZE, SUGAR, MALT FLAVOURING, SALT, NIACIN, IRON, VITAMIN B_6, RIBOFLAVIN (B_2), THIAMIN (B_1), FOLIC ACID, VITAMIN D, VITAMIN B_{12}.

NUTRITION INFORMATION

		Typical value per 100g	Per 30g Serving with 125ml of Semi-Skimmed Milk
ENERGY	kJ	1550	700*
	kcal	370	170
PROTEIN	g	7	6
CARBOHYDRATE	g	83	31
(of which sugars)	g	(8)	(9)
(starch)	g	(75)	(22)
FAT	g	0.7	2.5*
(of which saturates)	g	(0.2)	(1.5)
FIBRE	g	1.0	0.3
SODIUM	g	1.1	0.4
VITAMINS:		(%RDA)	(%RDA)
VITAMIN D	μg	4.2 (85)	1.3 (25)
THIAMIN (B_1)	mg	1.2 (85)	0.4 (30)
RIBOFLAVIN (B_2)	mg	1.3 (85)	0.6 (40)
NIACIN	mg	15 (85)	4.6 (25)
VITAMIN B_6	mg	1.7 (85)	0.6 (30)
FOLIC ACID	μg	333 (165)	110 (55)
VITAMIN B_{12}	μg	0.85 (85)	0.75 (75)
IRON	mg	7.9 (55)	2.4 (17)

It is sometimes important to know how much energy is stored in the food you eat. If you are an athlete, you may want a lot of extra energy. If you are on a diet, you may want to eat food with less stored energy.

The amount of energy stored in a food is called its **energy value**. The energy value of food is measured in **kilojoules**.

The packets of many kinds of food tell you the energy value of the food. The label tells you how many kilojoules (**kJ**) of energy are supplied when you eat 100 grams of the food.

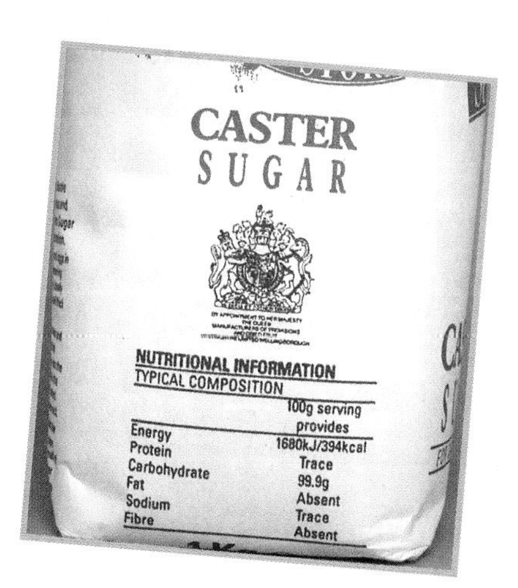

Energy for action

How much energy do you need from your food? That depends on who you are. A mountaineer climbing Mount Everest needs much more energy than a baby in a pram.

The table shows the energy values of some foods.

1 Which of these foods has the lowest energy value?

2 A bowl of cornflakes might have 50 g of cornflakes, 100 g of milk and 10 g of sugar. How many kilojoules is this?

Food	Energy value of 100 grams
cornflakes	1500 kJ
cooking oil	3700 kJ
peanuts	2400 kJ
bread	1000 kJ
margarine	3000 kJ
sugar	1700 kJ
milk	200 kJ

What do you know?

1 Choose words from this list to answer the questions that follow:

fuel **heat energy** **chemical energy** **food** **light energy**

a Name two energy stores.
b What form of energy is stored in a candle?
c What two forms of energy are produced when a candle burns?

2 We need energy from our food so that we can keep warm, and so that we can move around. The chemical energy from our food is changed to kinetic energy and heat energy. Draw an energy transfer diagram to show this.

3 a Draw a bar chart to show the energy values of the foods shown in the table on the left.
b Why are all the energy values given for the same amount of food (100 grams)?

4 a If you go on holiday to a hot country, you need to eat less food. Can you explain why?
b Children eat less than they used to 50 years ago. Can you suggest any reasons why children need less energy from their food today than they did in the past?
c A woman who is breast-feeding her baby needs to eat more food. Why do you think this is?

Key ideas

Food is the fuel that our bodies need for movement and to keep warm.

Foods have different **energy values**.

Energy values are measured in **kilojoules**.

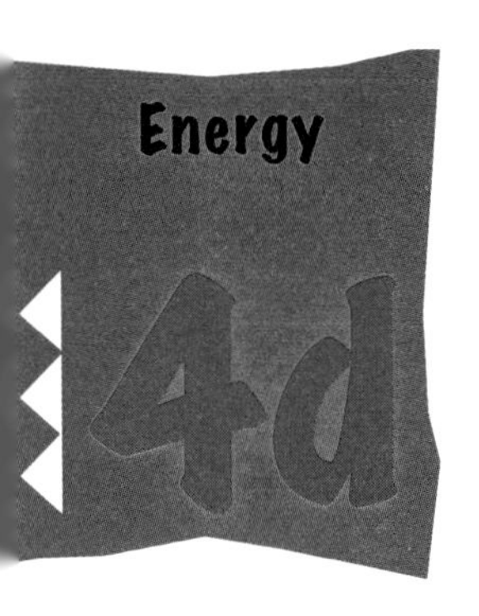

Electrical mysteries

Electricity is useful stuff. We use it for all sorts of different things.

a Can you name these mystery objects? They all need electricity to make them work.

Chemical stores

This toy electric train can run for hours on one battery. The battery is the store of energy that the train needs to move along the track. We know that a battery is a store of energy because it transfers its energy to the train, and the train moves.

Inside the battery are chemicals which store chemical energy. When the train is switched on, this energy is gradually changed to **electrical energy**. The train has an electric motor to make the wheels go round. Wires connect the battery to the motor. Electrical energy is transferred from the battery to the motor through these wires.

Here is an energy transfer diagram for the electric train.

chemical energy (of the battery) → electrical energy (in the wires) → kinetic energy (of the wheels)

Batteries have many different uses, because electrical energy can be changed into other types of energy. We couldn't have radios or televisions without electricity. A radio changes electrical energy into **sound energy**.

b What sort of energy is electrical energy changed into in the devices shown here?

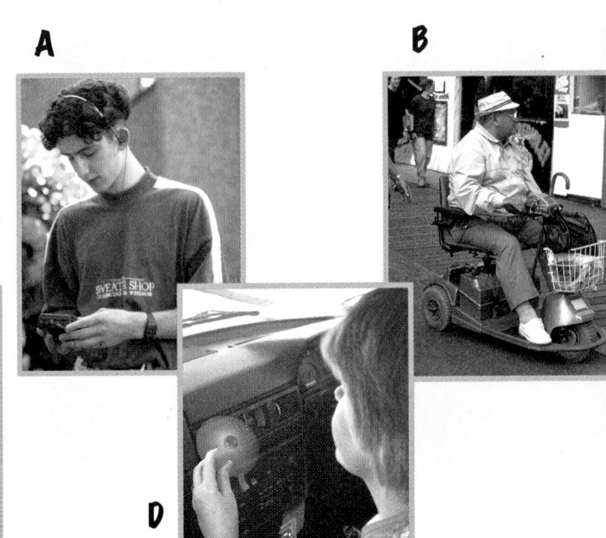

Generating electricity

Most of the electricity we use does not come from batteries. It comes into our houses along wires called the electricity mains. Where does it come from?

You can make electricity yourself if you have a **dynamo**. The girl in the picture is turning the handle of a dynamo to generate electricity. The harder she turns the handle, the more brightly the bulb glows.

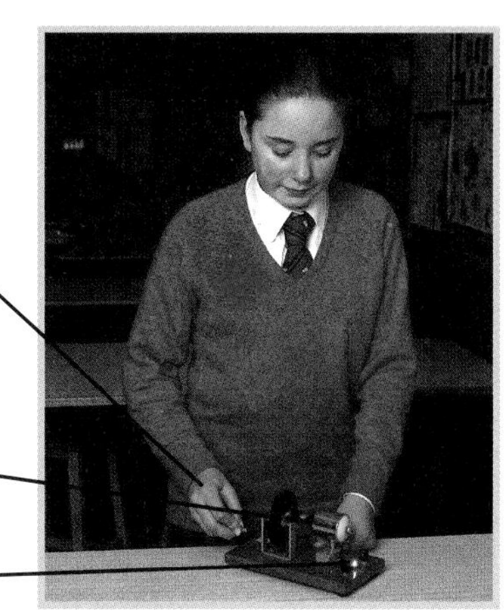

Most of our electricity comes from power stations where **generators** spin round just like the dynamo to generate electricity. The energy to turn the generators usually comes from burning coal or gas.

Some power stations work from the wind. This **wind turbine** is high up on a hill, where it is very windy. The blades of the turbine spin round as the wind pushes on them. There is a generator in the building down on the ground.

What do you know?

1 Look at the pictures of the mystery packages on the opposite page.
a What kind of energy do they all need to make them work?
b What kind of energy does each one produce?

2 Look at the picture of the wind turbine, and answer these questions.
a What kind of energy does the wind have? (Remember, wind is moving air.)
b What kind of energy does the spinning turbine have?
c What kind of energy does the generator produce?
d Draw an energy transfer diagram for the wind turbine.

3 Some bicycles have lights that work from batteries. Others have lights that work from a dynamo.
a Give at least one advantage of using batteries.
b Give at least one advantage of using a dynamo.

Key ideas

We can change **electrical energy** into several other kinds of energy.

Most of our electricity is generated in power stations that get their energy from burning fuels.

Hidden energy

When a light bulb is switched on, you see the light energy coming from it. You can hear the sound energy from a radio, and when you see something moving, you know that it has kinetic energy.

Some kinds of energy are harder to spot. Batteries store **chemical energy** which can be turned into electrical energy. Foods and fuels are also stores of chemical energy.

a All of the things shown here are stores of chemical energy. How could you use the energy they have stored in them?

Springy stores

Some toys make use of a different kind of stored energy. This car has a spring inside. When you wind it up, you transfer energy to the spring. The wound-up spring now has a store of **elastic energy**. When the car is released, the elastic energy in the spring is transferred to the car. The car moves along. It has kinetic energy.

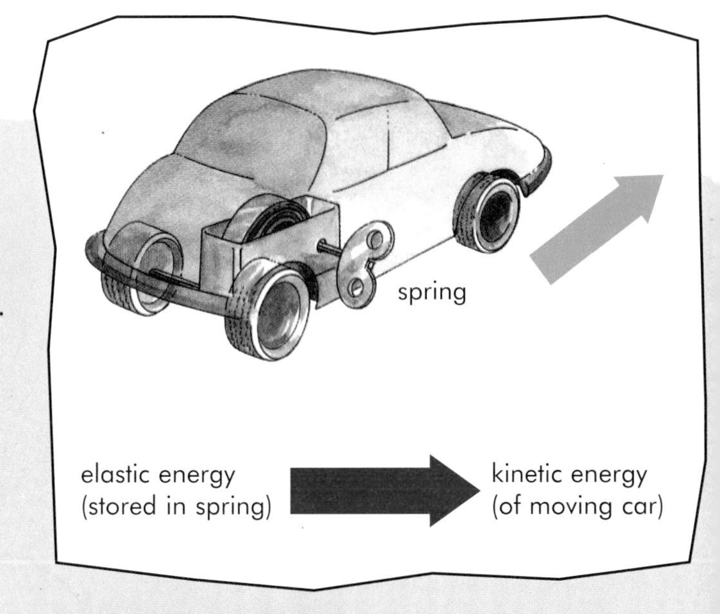

Rubber bands can also store elastic energy. Pull this 'jumper' down to stretch the rubber band. Let go, and it jumps up from the ground. The more the rubber band is stretched, the greater the amount of elastic energy it stores, and the higher it jumps.

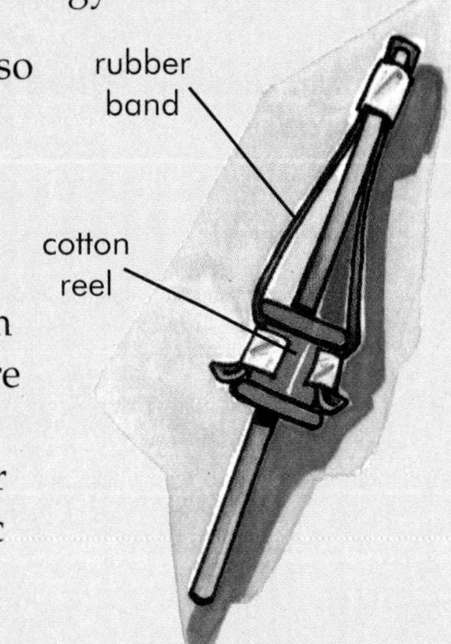

Elastic energy is important in lots of sports. The picture shows an example.

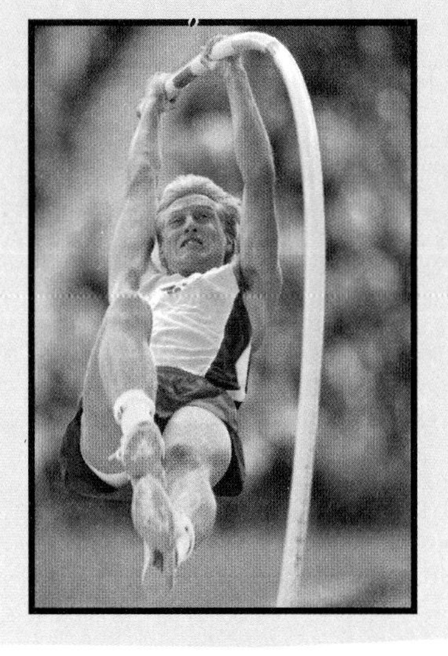

The pole vaulter sticks the end of his pole into the ground. It gets bent, and then lifts him up into the air as it straightens out.

Gravitational energy

You need energy to lift things up. A heavy object is difficult to lift because of the force of gravity pulling it down. When you have lifted it up, it has **gravitational energy**. This is another kind of stored energy. You know that it has energy because if you let it go, it can fall and hurt your foot!

Anything that is high up has gravitational energy. We can make use of that energy to do things for us.

There is a lot of water behind this high dam. It stores a lot of gravitational energy. When the water is released, it flows down inside giant pipes and turns an electrical generator.

Gravitational energy has been changed to kinetic energy and then to electrical energy. This is called a hydroelectric power scheme.

What do you know?

1 This table shows some examples of stores of energy. Unfortunately, many of the examples are in the wrong columns. Copy the table, and put each example in the correct column.

Chemical energy	Elastic energy	Gravitational energy
a stretched spring	a battery	fuel
a raised hammer	food	water behind a dam
a squashed rubber ball	a skier at the top of a slope	a stretched rubber band

2 Draw an energy transfer diagram to show the energy transfers for a hydroelectric power scheme.

3 a Why is elastic energy important in these sports and games?

diving off a springboard **riding on a pogo stick** **bungee jumping**

b A gymnast jumps up and down on a trampoline. Each time, he presses the rubber sheet down a little further. How is the elastic energy stored in the rubber changing?

Key ideas

Gravitational energy is stored by anything that is lifted off the ground.

Elastic energy is stored in anything that is stretched or squashed.

Chemical energy is a third type of stored energy.

EXTRAS

4b Barbecueing yourself

If you have ever helped with a barbecue, you will know that you have to be careful not to burn yourself. Some of the heat energy from the burning charcoal goes into the food, but a lot of it spreads out around the barbecue.

1 What happens to the heat that does not go into the food?

2 Draw an energy transfer diagram showing what happens to the heat coming from a barbecue.

3 Why is it very wasteful to use a barbecue to cook just one sausage?

4 How would you make the best use of the heat from the barbecue, so that as little as possible was wasted?

Fuel to go

Fuels are useful for travelling. A car carries an energy store of petrol or diesel fuel. The fuel is burned in the engine, and the car moves along. The chemical energy of the fuel has become kinetic energy of the car.

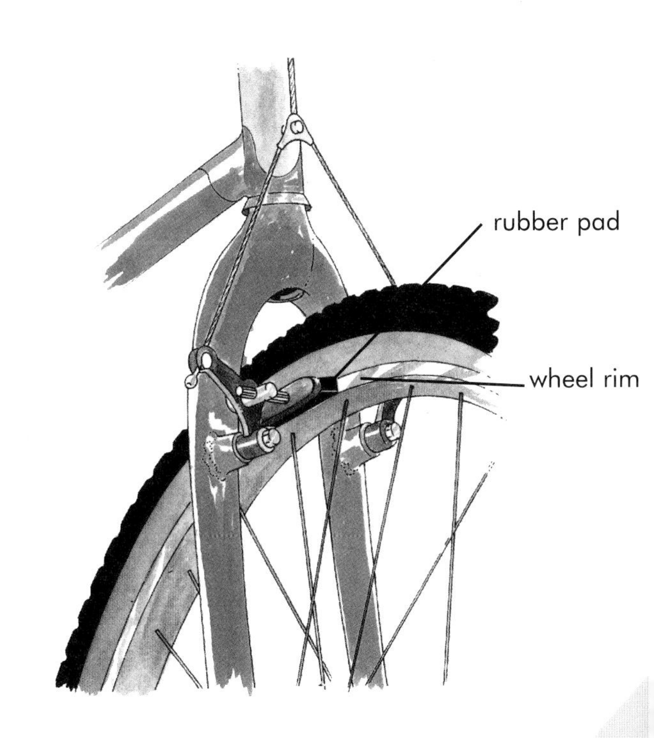

5 Draw an energy transfer diagram to show how energy is transferred from the fuel to the car when the car is travelling along.

6 When a cyclist slows down, the brakes rub on the rim of the wheel. The rubber pads get hot. Draw an energy transfer diagram to show this energy transfer.

4d The great escape

This picture shows a steam engine, generating electricity. Unfortunately, a lot of energy is escaping.

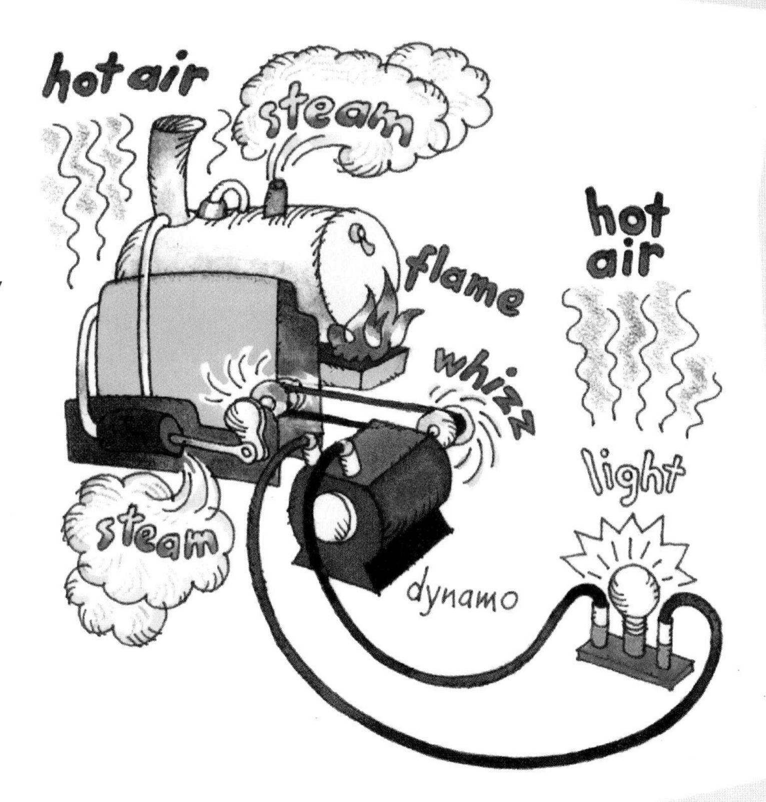

1 Make a list of the different places where energy is escaping, and the type of energy that is escaping at each place.

2 The engine uses paraffin as its fuel. Is paraffin a type of energy?

3 Steam is escaping from the engine. Is steam a type of energy?

4 Draw an energy transfer diagram for the complete system.

5 Which is bigger, the amount of energy in the fuel at the start, or the amount of electrical energy that the dynamo supplies to the lamp?

4e Energy for clocks

Clocks need a store of energy to keep them going.

1 Look at the clocks shown here. What sort of energy does each one use?

a a battery-operated clock

b a clockwork clock

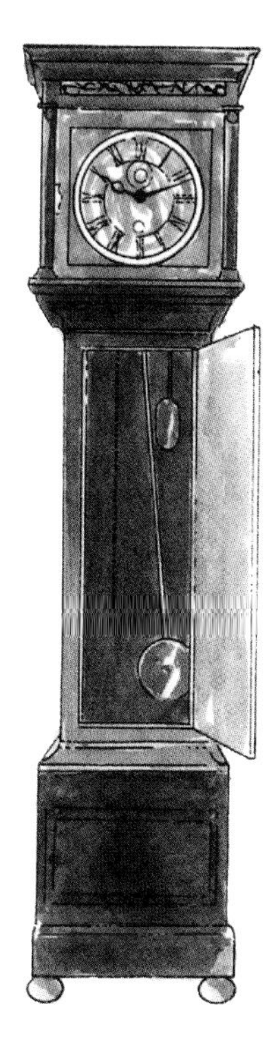

c a grandfather clock with a pendulum and a weight

5a Sun for supper

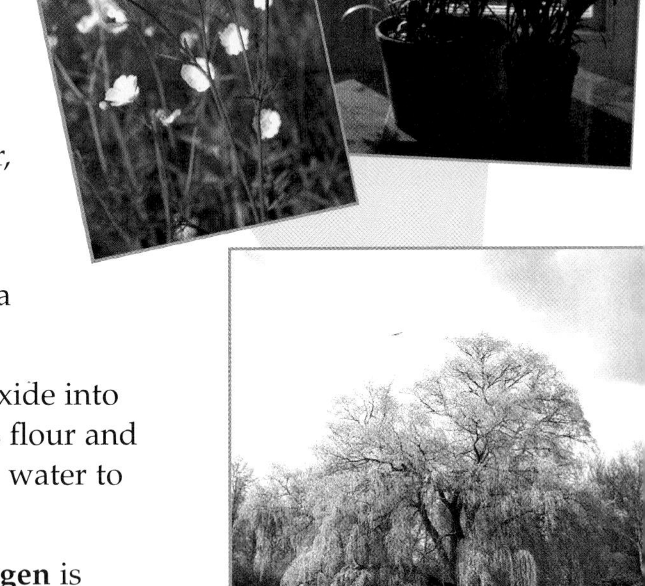

a Think about the plants you have seen at home or at school. What do they need to keep them healthy and make them grow?

Food for thought

All living things need food for energy, but plants don't eat food. Plants need three main things, which are water, light and air. Plants use these to make their own food.

The main ingredients for the plant's food are **water** and **carbon dioxide**, a gas in the air. Carbon dioxide is only a small part of the air, but it is very important to a plant.

Light provides the **energy** to turn water and carbon dioxide into food, a bit like the way the heat energy in an oven turns flour and water into bread. The plant uses the carbon dioxide and water to make food substances called **carbohydrates**.

When carbohydrates are formed, another gas called **oxygen** is produced. This is the waste product of the process.

Making food using energy from the sun is called **photosynthesis**.

The leaf factory

Photosynthesis happens in the **leaves** of plants. Water enters the plant through its roots and travels up the stem to the leaves. Carbon dioxide enters through tiny holes in the leaf.

Light energy hits the leaf. The **chloroplasts** trap the light energy so they can use it to make food. Carbohydrates are made in the leaf and transported all over the plant to be used for growth. Oxygen escapes through the tiny holes in the leaf's surface.

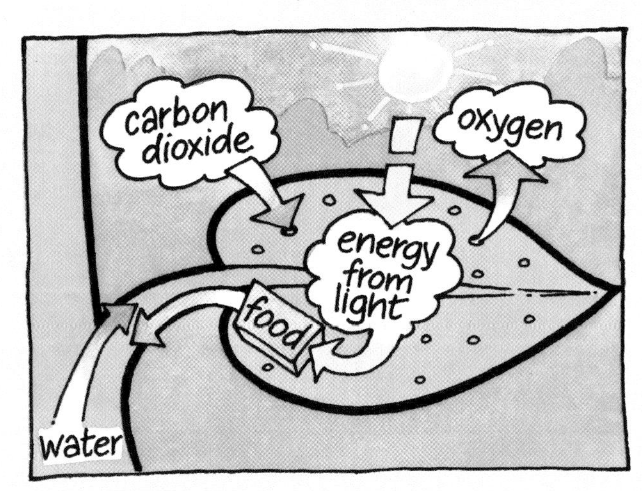

The photosynthesis production line

Light for life

Imagine lying on your back underneath a tree on a warm, sunny day. The view you see is just like this picture. The bright green leaves are arranged so that little or no light falls onto the ground underneath where you are resting.

Why do you think the leaves grow like this?

Not all plants are completely green. Some have leaves that are partly white or yellow. This looks very pretty, but it can cause problems if the plant does not get plenty of light. In poor light the leaves often turn completely green again, or the plant may die.

Can you work out why?

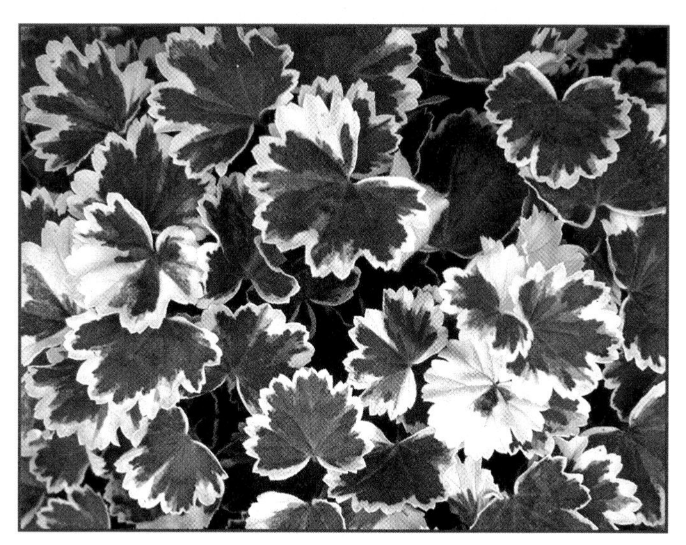

What do you know?

1 Copy and complete the following sentences. Use the words below to fill the gaps.

leaves carbohydrates carbon dioxide photosynthesis oxygen water light

Plants make their own food in their ________ using ________, ________ ________ and energy from ________. The way plants make food is called ________. They produce ________ and ________.

2 What would happen if you put a plant into a dark cupboard and left it for several weeks?

3 People breathe in oxygen and breathe out carbon dioxide. Some people claim that talking to house plants makes plants grow better. Can you think of a scientific explanation for this?

Key ideas

Plants make food by **photosynthesis**.

In photosynthesis, plants use **light energy** to turn **water** and **carbon dioxide** into **carbohydrates** and **oxygen**.

Photosynthesis takes place in the **leaves** of a plant.

5b The root of it all

The leaves of plants make food (they photosynthesise) and so they are very important. But most plants also have parts that are not green and that you hardly ever see. These parts are just as important as the leaves. They are the **roots**. Without their roots, plants cannot survive.

a Make a list of the jobs you think plant roots might do.

Holding on

Plants need to be held firmly in one place, so that they can spread their stems and leaves to trap as much sunlight as possible, and so that the wind does not blow them over.

The roots spread underground and act as an anchor for both the plant and the soil. Without plant roots, the soil is blown away and only bare rock is left.

Water, water!

Plants need water for photosynthesis to make their food. They also need water to help them move substances around the plant.

How do plants get the water they need? The answer is through the roots.

As roots grow they become covered in tiny root hairs. Water enters the plant through these root hairs. Water moves from the soil into the root hair cells.

From here the water moves through the root and into the plant itself. The plant's transport system carries the water all around the plant to where it is needed.

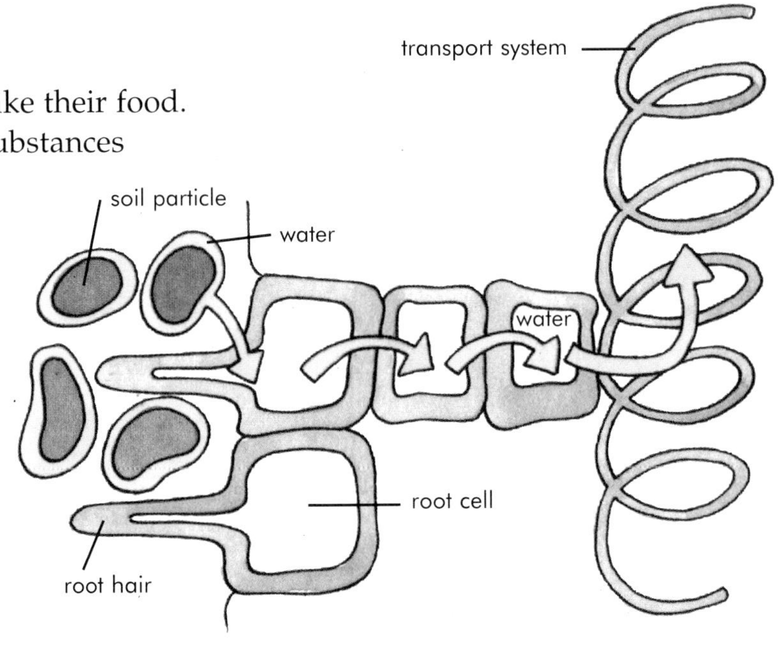

Plant food from the soil

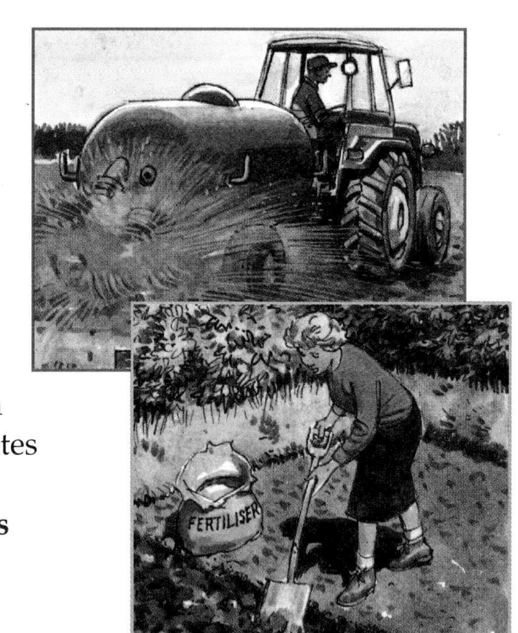

To make sure their crops grow well, farmers regularly put **fertiliser** on their fields. The fertiliser may be manure from animals, or special chemicals that the farmer can buy.

Gardeners fertilise their soil to keep their vegetables growing large and their flowers blooming.

Plants make their own food, so why does everybody give them fertiliser? When plants photosynthesise, they make carbohydrates out of carbon dioxide and water. To be healthy and to grow, plants need other things too, particularly **nitrogen**, **phosphorus** and **potassium**. These all come from well fertilised soil.

Why fertiliser?

Here are some clues about why plants need other things from soil. Use them to describe why plants need:

1 nitrogen
2 phosphorus
3 potassium.

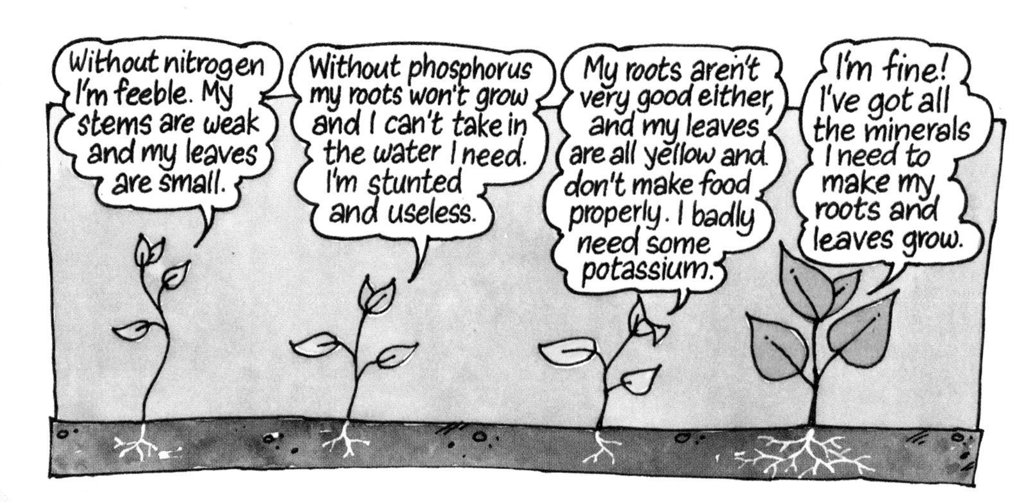

What do you know?

1 The roots of a plant have three main jobs. What are they?

2 Make a small leaflet or a poster for your local garden centre. It should explain to people why they need to buy fertiliser to feed their plants. Make it bright and cheerful so that people want to look at it, with clear explanations so they understand it.

3 The label on a bottle of house plant food claims that two drops in the water will help house plants to grow larger and better. Plan an investigation (which must be a fair test) to show whether this is true.

Key ideas

The **roots** of a plant hold it firm in the soil, and take up water and other substances that the plant needs to make its food.

People add **fertiliser** to soil to help plants grow well.

Nitrogen, **phosphorus** and **potassium** are important ingredients in fertiliser.

5c Animals need food too

Animals cannot make their own food. To get the energy they need to live and grow, they must eat plants or other animals.

a Jot down as many animals as you can, along with what they eat.

Not too much, not too little

Animals, including people, need to eat just the right amount. If animals eat more food than they need for living and growing, the extra becomes a layer of fat. This acts as an energy store, but too much fat is very unhealthy.

If animals don't get enough to eat, they get very thin and catch diseases easily. If they have very little or no food for long periods, they die.

Eating the right things

Like plants, we need a lot of different substances in our food. These are some of the main ones.

- **Carbohydrates** are full of energy which the body can use very easily.
- The body uses **protein** to grow and to repair itself.
- **Fat** has much more energy than carbohydrate. We need some fat to keep us healthy, but too much can make us overweight and give us heart problems.
- We only need tiny amounts of **vitamins** and **minerals**, but they are vital in keeping us healthy.

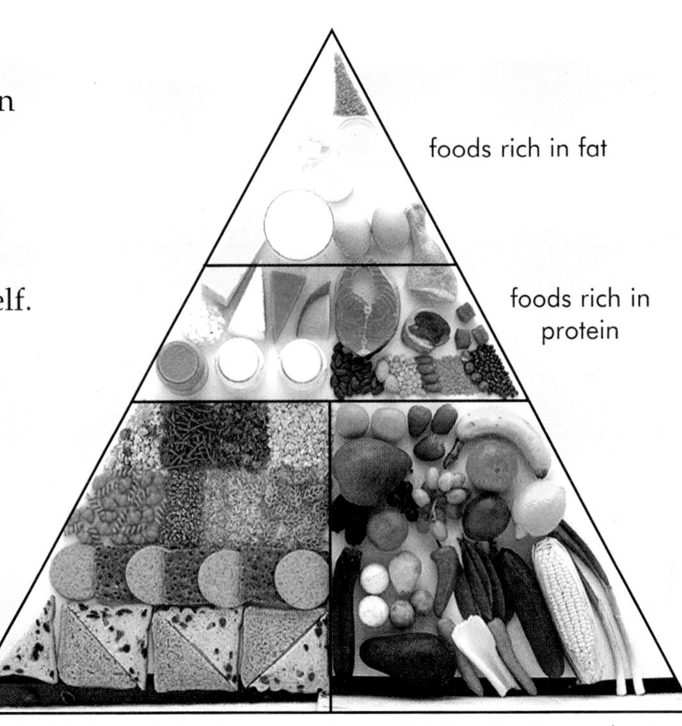

Here are the sorts of foods that Thomas and Claire have every day. Are they eating all the things they need?

1 Work out what sort of foods they don't eat enough of.
2 Suggest a healthier diet for each of them.

Thomas

Sunday: Breakfast: Got up late – didn't bother. Cup of coffee with three sugars
Lunch: Roast beef, roast potatoes, roast parsnips, Yorkshire puddings, gravy, peas. Apple pie and cream
Tea: Buttered toast, trifle, bit of Gran's chocolate cake

Monday: Breakfast: nearly late for school – grabbed a Kit-Kat
Lunch: Pizza and chips, shortbread, coke
Tea: Sausages, chips and beans, ice-cream

Tuesday: Breakfast: Football practice this morning – got up in time for a bacon butty
Lunch: Chicken nuggets with chips and salad, doughnut
Tea: Fish and chips, mango cream

Claire

Thursday: Breakfast: Muesli with skimmed milk, fruit juice
Lunch: Cottage cheese salad, apple
Tea: Baked potato, green beans, carrots and tomato. Fruit salad

Friday: Breakfast: Porridge, spring water
Lunch: Two dry crispbreads, apple, orange
Tea: Poached fish, small boiled potato and salad, fresh pineapple

Saturday: Breakfast: All-bran with fruit juice, prunes
Lunch: Raw carrot, banana and apple
Tea: Chicken casserole, broccoli and carrots, didn't bother with pudding – Mum had made one of her rich fruit pies with cream!

A balanced diet

We need all the food substances in the right amounts. When a diet has the right things in the right amounts, it is called a **balanced diet**.

What do you know?

1 Copy and complete these sentences. Use the words below to fill the gaps.

fat food vitamins balanced store minerals proteins

Human beings and other animals must eat ________ to get energy. If we eat too much we get ________ as our bodies ________ the energy in the food. To be healthy, we need to eat a ________ diet with the right amounts of fat, ________, carbohydrates, ________ and ________.

2 Explain each of these facts:
a Children need more protein in their diet than adults.
b When people try to lose weight, they need to cut down the amount of fatty food they eat.

3 Why do you never see a fat plant?

Key ideas

Animals cannot make their own food. They eat plants or other animals.

The important parts of a person's diet are **carbohydrates**, **fat**, **protein**, **vitamins** and **minerals**.

These must be eaten in the right amounts for a healthy, **balanced diet**.

5d Healthy eating

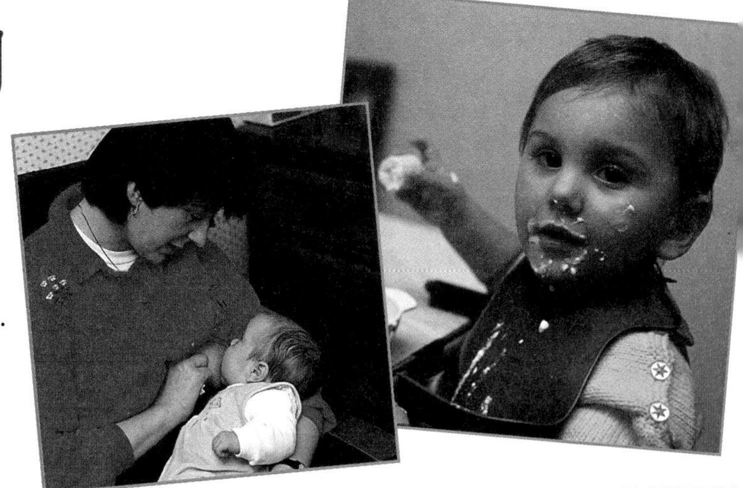

All that small babies need as food and drink is milk, from their mothers or from a bottle. But as babies get older milk is no longer enough. By about a year old babies are eating solid food. Why do they need solid food?

Minerals and vitamins

Milk gives us the proteins, carbohydrates and fats that we need, but it does not have all the **minerals** and **vitamins** we need to grow. We only need tiny amounts of these but without them we get very ill and our bodies don't work properly. Some of the vitamins and minerals we need to eat are described below.

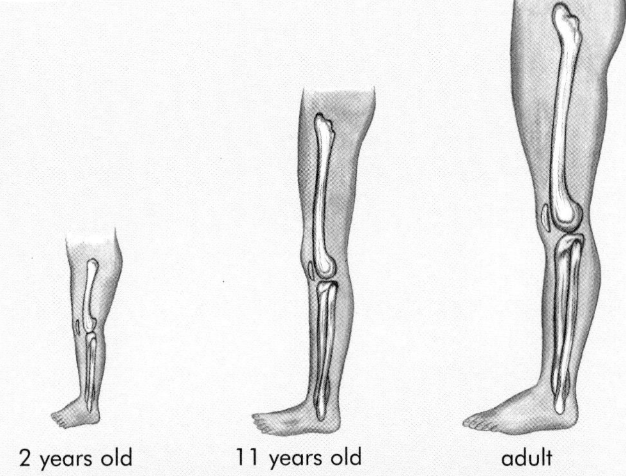

As we grow up, our bones get bigger and longer. We need the mineral **calcium** to make them hard and strong. We get a lot of calcium when we drink milk or eat dairy products.

We can run fast because our blood carries lots of oxygen to our cells. We need the mineral **iron** for our blood to carry oxygen properly. Without it we feel tired and weak. We get iron when we eat red meat or apricots.

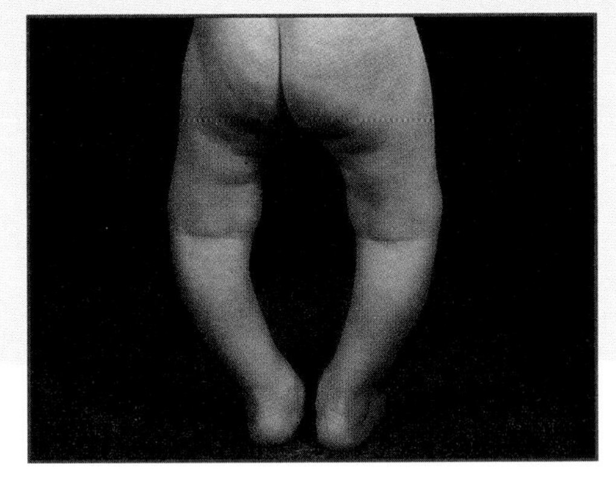

We need **vitamin D** as well as calcium to make sure our bones and teeth grow strong. If we don't get enough vitamin D, our bones are soft and will not hold our weight without bending. We get rickets like this little girl. Cod liver oil is full of vitamin D.

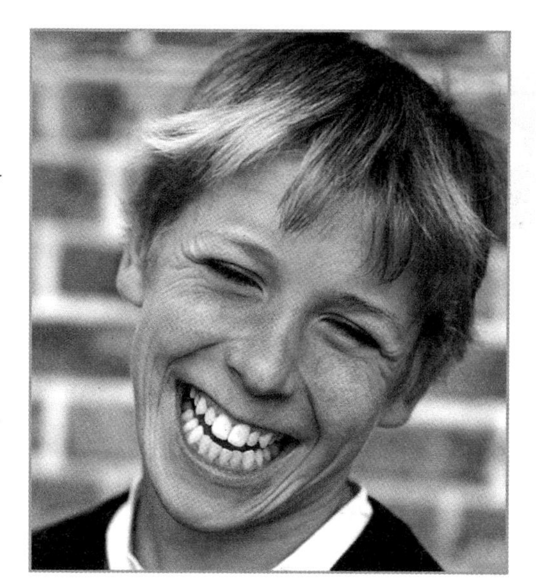

Vitamin C is found in oranges and lemons and lots of green vegetables. We need it for healthy gums and skin. In the olden days sailors didn't get fresh fruit on long sea voyages, and many died of scurvy, the disease caused by lack of vitamin C.

1 Make a table with the headings iron, calcium, vitamin C and vitamin D.

2 Write down foods containing these vitamins and minerals under the headings and say why they are important in the body.

3 If you can, find out about some more vitamins and minerals and add them to your table.

Regular habits

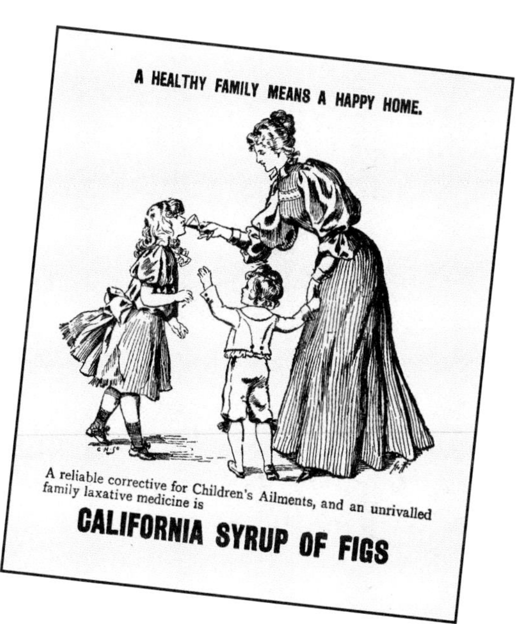

The food we can't use should be pushed out of our bodies as faeces within 24 hours. If our food doesn't move through the body quickly, the result is constipation. Years ago children were given horrible potions to make sure they went to the toilet regularly!

Now we know there is a much healthier way of keeping things moving. We don't use all of the food we eat, for example, the cellulose from plant cells. We call this material **fibre**, and it helps to keep the food moving through the body. Fruit, wholemeal bread, bran, beans and sweetcorn have a lot of fibre.

What do you know?

1 Copy and complete the following sentences. Use the words below to fill the gaps.

fibre **vitamins** **body** **blood**
minerals **tiny** **calcium**

_______ and _______ are substances which our bodies need in _______ amounts to be able to work properly. Milk is a good source of _______ for healthy bones and teeth, but it does not contain the iron the _______ needs to carry oxygen properly. _______ is food material which we cannot use. It keeps food moving through the _______ regularly.

2 Your company has just produced a new health tablet containing a mineral or a vitamin. Choose a vitamin or mineral from the four on these two pages, and then design either a leaflet or a poster advertising your product. It must explain to people what the product is and why they need it.

Key ideas

Tiny amounts of **minerals** and **vitamins** are needed in the diet for a healthy body.

Fibre is needed in the diet to keep food moving through the gut properly.

5e Food for thought

You can see food like this every day in canteens, restaurants and at home. But where does all the food come from?

Much of this meal comes straight from plants. The bits that come from animals, the meat, butter and milk, also come from plants, but less directly. The cow that produced the meat and also milk to make the butter ate grass. In the end, all the food comes from plants.

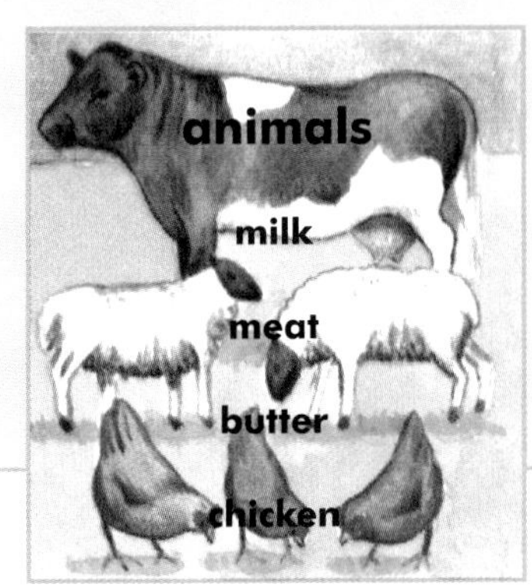

Where does your food come from?

a Write down the food you ate for your last three meals. Now put each meal into boxes, the plant food in one box and the animal food in another. Then link all the animal food to plants.

What about animals?

It is easy to see that this leaf insect is getting its energy from plants. What about the cat and the killer whale? The cat is eating a mouse, but the mouse ate grass seeds and berries, which come from plants. Even the killer whale eventually depends on plants. The whale eats the seal, the seal eats fish, they eat smaller fish, and the little fish eat the tiny plants that grow in the sea.

The producers

All animals depend on plants for food. Why? Because only plants can make new food by photosynthesis. They use energy from the sun to make carbohydrates, and then they use these for living and growing. Only plants can do this, and so they are known as the **producers**. So all living things depend on the energy of the sun.

We're the consumers

Plants are the producers and animals are **consumers**. Animals eat plants, or they eat animals that eat plants.

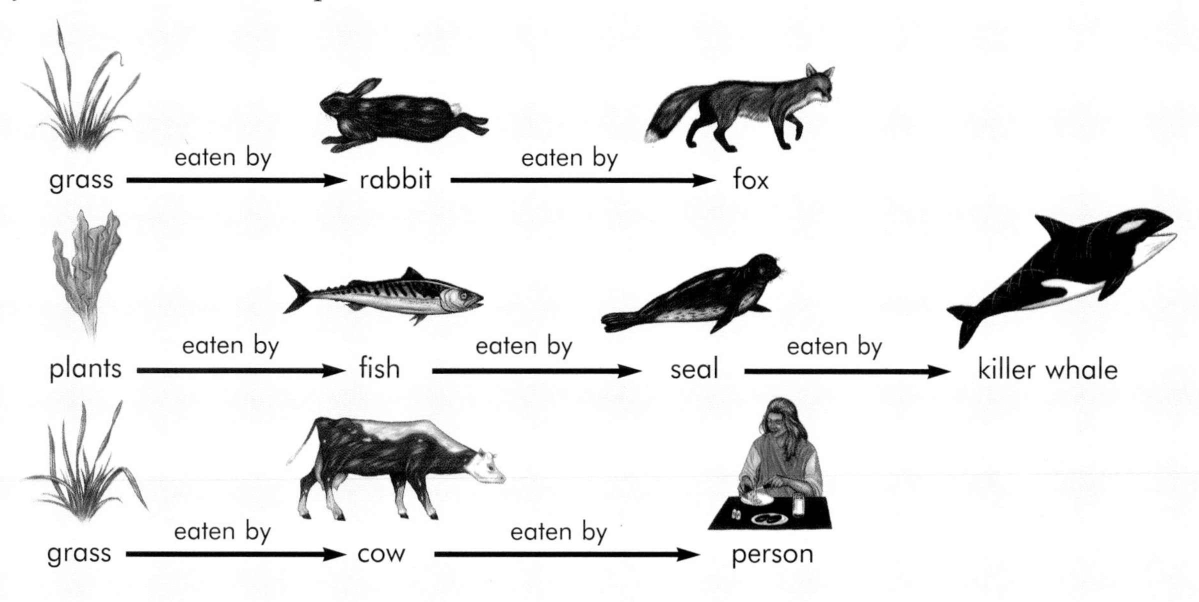

Plants and animals are like the links in a chain, which can be followed back to the energy from the sun. When we link organisms like this, we are looking at **food chains**.

What do you know?

1 a Why are plants called producers?
b What are consumers?

2 Design a poster called 'The chains of life' for the science area of your school.

3 Why do food chains show that all life depends on the energy from the sun?

4 Think up and draw out three more food chains of your own.

Key ideas

Plants are **producers**. They produce new food from carbon dioxide and water using energy from the sun.

Consumers are living organisms that need to eat other organisms to get their energy and to grow.

Food chains are the links between different animals that feed on each other and on plants.

5f The web of life

Pond links

In any **habitat** plants and animals are linked together by the sort of food chains we have already seen. But if we look a little closer, we can see that things get a bit more complicated.

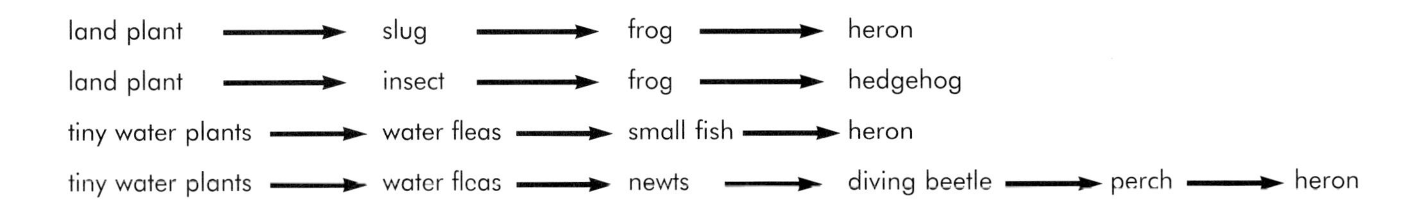

Water fleas are eaten by fish and by newts and by diving beetles. Frogs are food for hedgehogs, foxes and herons. How can we show this?

We can use a **food web**. This connects all the different organisms that live together and feed off each other. Here is a food web for the edge of our pond.

Try to make another.

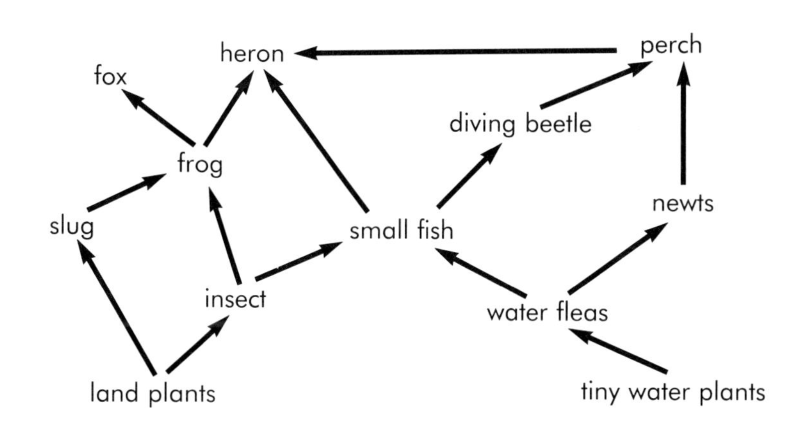

Webs in the cold

Here are some food chains that can be found in the Antarctic. Use them to make a food web.

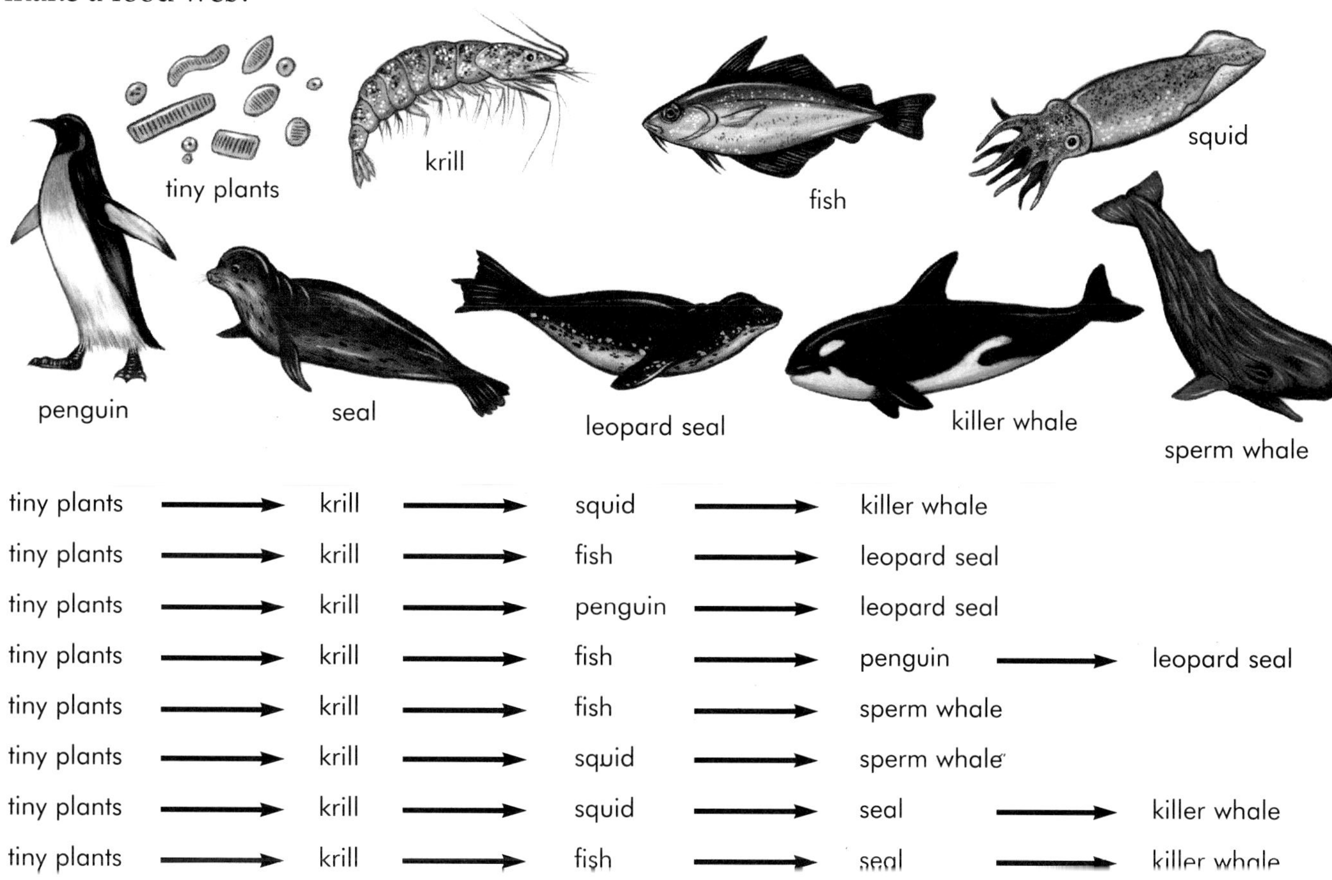

What do you know?

1 Look at the pond food web. Make a list of the organisms that are the producers, and another list of the organisms that are the consumers.

2 Explain why a food web is a much better model of what really goes on in the living world than a food chain.

Key ideas

The **habitat** of an animal or plant is the place where it lives.

A **food web** is a model of a habitat showing how the animals and plants in it are connected.

5 EXTRAS

5a Who grows tallest?

Joe Tumbleweed is a gardener who has a big greenhouse. He uses artificial lights at night in the greenhouse. He changed the number of hours of light he gave to his cress seedlings and measured them to see if they grew by different amounts. Here are his results.

Hours of light	Average growth of seedlings in 24 hours (mm)
10	4
12	6
14	8
16	10

1 Make a bar chart of Joe's results.

2 Which plants have grown the tallest?

3 Explain what has happened to the cress seedlings.

5c Slimmer of the year

In many parts of the world, eating a balanced diet is difficult because food is in short supply. In other parts of the world, there is plenty of food but some people eat badly balanced diets. They eat more food than they need for energy (see Unit 4) and so they get fat.

1 How do people get fat?

2 This woman has clearly lost a lot of weight.
a What sort of changes would she need to make to her diet to lose all this weight?
b What other changes could she have made to her way of life to help her lose weight?

3 Most people in the United Kingdom are not seriously overweight. But many people, particularly women, follow 'slimming diets'. Why do you think they are so concerned about how much they weigh?

5d In at one end, out at the other!

In many parts of the world, people eat a diet which is made up largely of plant material. They eat a lot of pulses, fruit and vegetables, but only a little meat. Scientists have collected the waste produced by people eating in this way and found that they produce about 1 kilogram of soft faeces every day. In other parts of the world people eat a lot of meat and much less plant material. They produce about 200 grams of hard faeces a day.

1 Explain this difference.

2 It is healthy for waste from any particular meal to pass through the body in about 24 hours. Which type of diet is most likely to make this possible?

3 What sorts of waste would you expect the following animals to produce and why?

elephant	**lion**	**rabbit**	
zebra	**cow**	**owl**	**fox**

5f Woodland webs

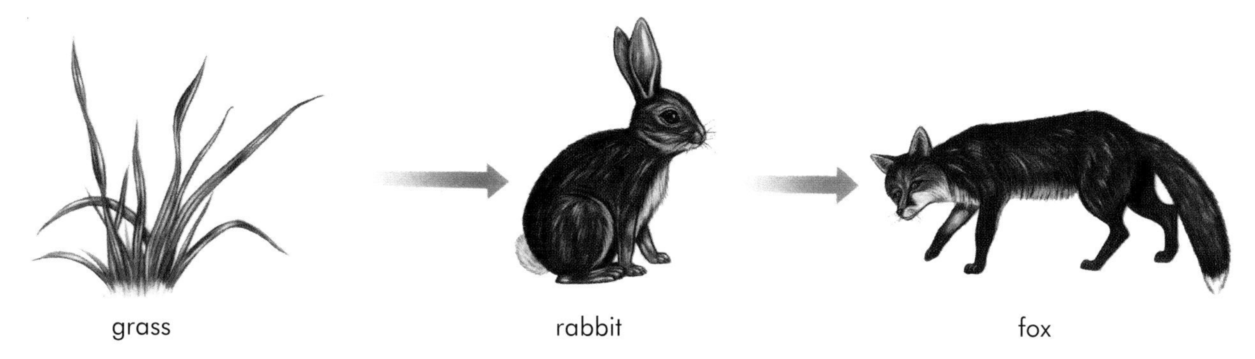

This is a simple woodland food chain.

1 If the summer is hot and dry and the grass dies off, what will be the effect on the rabbits and the foxes?

2 If a disease attacks the rabbits, what will be the effect on the grass and the foxes?

3 Here are some woodland animals and plants.

grass	**woodland plants**		**rabbits**	**mice**
voles	**beetles**	**owls**	**foxes**	**badgers**

Use them to produce a woodland food web.

4 A web is a much better model than a chain of what happens in the real world. How do you think a change in the amount of grass would affect the rest of the web? What about a change in the number of rabbits?

6a Separating mixtures

a How do you go about separating a mixture?
Decide how you would separate:
- the red Smarties
- the carrots from the soup
- the different coins.

Sorting by size

If you had a mixture of peas and sand, you could pick the peas out one by one, but that would take ages. There is a faster way to separate them. Peas are much larger than grains of sand, so you could put the mixture through a **sieve**. This lets the sand through but not the peas.

Other types of sorting

To separate mixtures, you need to find a difference between the mixed substances which you can use to separate them out.

b Use the information to work out how to separate:

1	**snooker balls**	**ping-pong balls**
	size about 4 cm	size about 4 cm
	sink in water	float in water
	various colours	white
2	**iron filings**	**copper filings**
	grey	brown
	sink in water	sink in water
	magnetic	non-magnetic
3	**aniseed balls**	**Polo mints**
	brown	white
	round	flat
	sink in water	sink in water

What a mix-up!

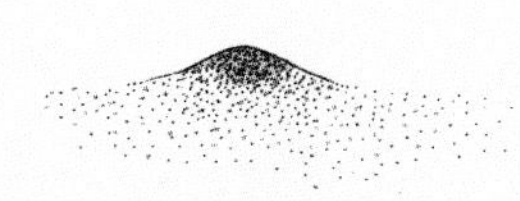

Helpful hints

Solids from liquids: decanting

Has anyone in your family ever made home-made wine? You start off with a mixture of fruit and yeast in water, and end up with a bubbling mess.

But soon most of the solid material settles out at the bottom of the bottle. This is called **sediment**. If you are careful, you can pour the clear wine off, leaving the solid behind. This method of separation is called **decanting**.

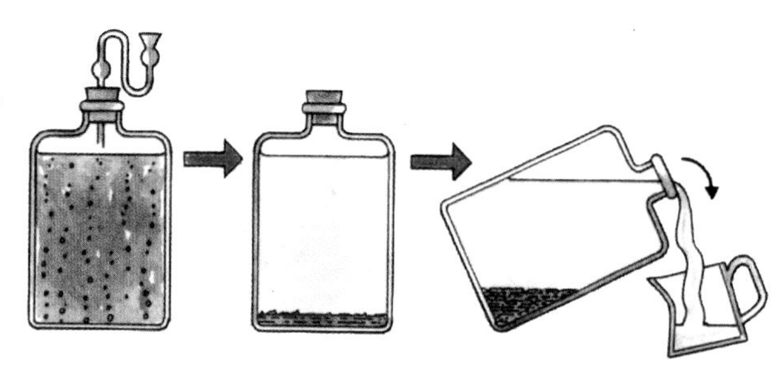

Solids from liquids: filtration

After the wine has been decanted, it is often still cloudy. There are still some small pieces of solid left mixed up with the liquid. These are in **suspension**. They can be separated out using filter paper. This is a sieve with very tiny holes. The liquid wine can get through the holes, but any solids in suspension are trapped.

This process is called **filtration**. The solid left behind in the filter paper is the **residue**. The clear liquid in the jar is the **filtrate**.

The filter paper is folded into a cone and put into the funnel.

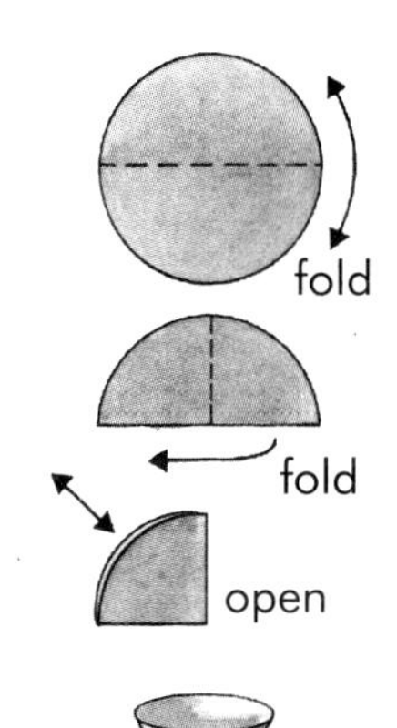

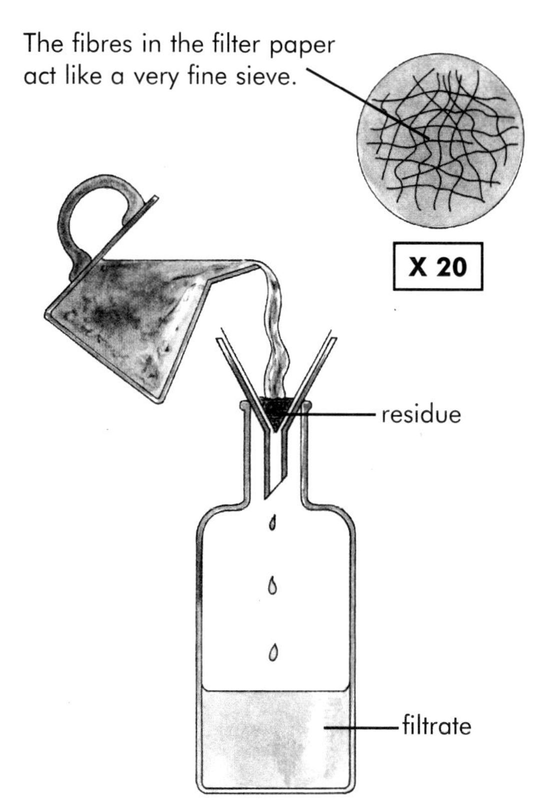

What do you know?

1 Peas are green and sand grains are yellow-brown. Why can't you use this difference to separate peas and sand?

2 A car engine will not work properly if dust or dirt gets mixed up with the petrol. The petrol going to the engine passes through a container filled with matted fibres.

a What is this for?

b How does it work?

Key ideas

Solids of different sizes can often be separated through a **sieve**.

If a mixture of liquid and solid settles out, the liquid can be **decanted** off.

Solids can be **filtered** out of liquids.

6b Solids in liquids

Fruit trees and grape vines are often sprayed with a special mixture to protect them from disease. Gardeners make this spray by mixing a substance called copper sulphate with water and a little lime.

Solid copper sulphate comes in beautiful blue crystals. When these crystals are mixed with water, they seem to disappear. But the water turns blue, so copper sulphate must still be there. What has happened to it?

Dissolve it . . . slowly!

Copper sulphate **dissolves** in water and makes a **solution**. This makes it easy to spray onto the vine leaves.

If you drop some copper sulphate crystals into water and leave them alone, you can watch them dissolve (if you are very patient). The crystals slowly disappear and the water turns blue. It would take many hours for all the solid to dissolve.

Speed it up

Fortunately, you can make things dissolve faster. There are three main ways of speeding the process up.

1 Heat it up. Heat makes solids dissolve faster.

2 Shake or stir the mixture. This mixes the solid with the liquid faster.

3 Crush it. Small pieces of solid dissolve faster than large ones.

More about dissolving

Substances that dissolve in water to form a solution are described as **soluble**. The liquid is called the **solvent** and the solid is called the **solute**. Copper sulphate is soluble in water. Water is the solvent and copper sulphate is the solute.

How much copper sulphate could you dissolve in water? Could you go on adding more and more, making the solution stronger and stronger?

The answer is no. You would eventually find that no more copper sulphate would dissolve, no matter how long you stirred the mixture. When this happens, the solution is **saturated**.

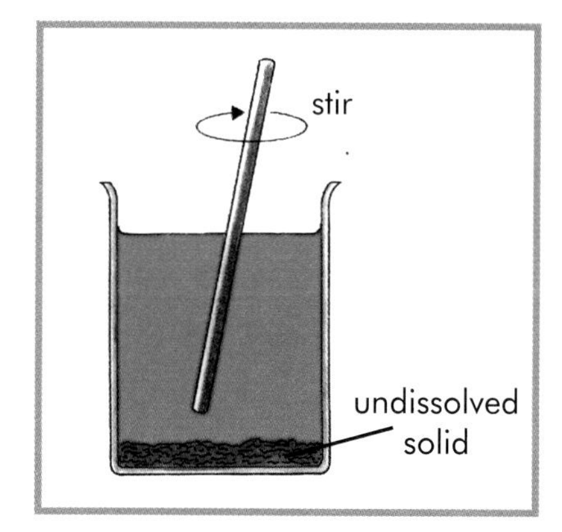

A saturated solution

What dissolves?

Not everything dissolves in water. A lot of things, such as chalk and oil, will not dissolve. They are **insoluble**.

Think carefully and decide if the following solids are soluble in water or not:

sand flour washing powder salt sugar
glass plastic cups paper instant coffee

What do you know?

1 Copy out the following descriptions and match the correct word to its description. One has been done for you.

the solid that dissolves	**insoluble**
the liquid that does the dissolving	**solute**
it's full up with solid	**solution**
it will dissolve	**solvent**
it will not dissolve at all	**soluble**
a liquid with solid dissolved in it	**saturated**

2 There are three things you can do to speed up dissolving. What are they?

3 Some breakfast cereals have a frosting of fine grains of sugar. Why do these cereals taste sweeter than ordinary cornflakes that have been sprinkled with granulated sugar?

Key ideas

Some solids **dissolve** in water to form a **solution**.

The water is the **solvent**, and the solid is the **solute**.

If no more solid will dissolve, the solution is **saturated**.

You can speed up dissolving by crushing the solid or heating or stirring the liquid.

Solids that will not dissolve are **insoluble**.

6c Separating solutions

When it rains, everything gets wet. But if it stops raining and the sun comes out, everything dries up again. Where does all the water go?

Evaporating away

Water can slowly turn into a gas at any temperature. If you leave water in an open bowl at home, some of the water turns to a gas called **water vapour** and mixes with the air. This process is called **evaporation**.

Evaporation dries up puddles, and dries out clothes on the line. The hotter it is, the faster evaporation happens.

Useful evaporation

When salt dissolves in water, the salt particles are far too small to be filtered out by filter paper. The sea contains salt in solution. How can you get the salt out?

The simplest way is to let the water evaporate away. The salt particles join back together as the water is lost, and crystals start to grow. It's like dissolving in reverse.

This method has been used to get salt from sea water for thousands of years. Sea water runs into shallow 'salt pans', which are then blocked off from the sea. The sun evaporates the water, leaving the salt behind. 'Salt pans' are still used around the Mediterranean Sea today.

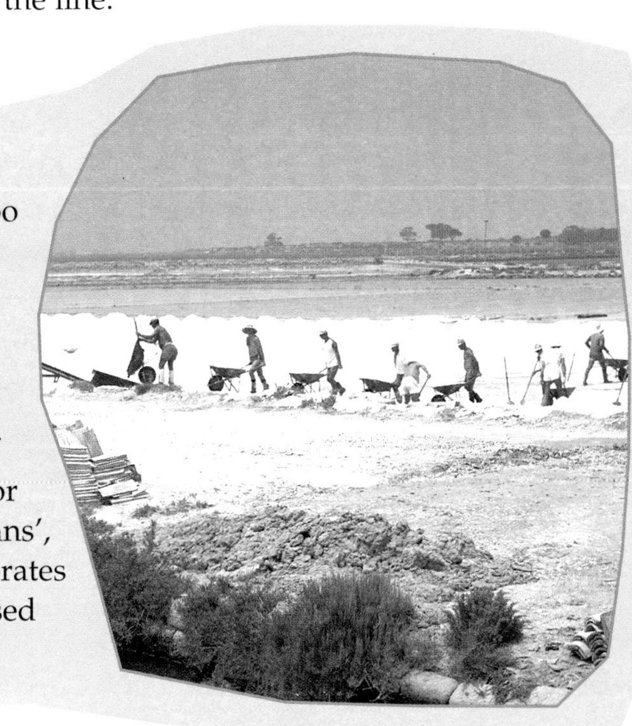

Pure salt from the salt mines

Most of the salt used today comes from underground mines. The rock salt that is dug up is mixed up with sand and clay. How can it be separated?

A useful difference between salt, sand and clay is that only the salt is soluble in water. This difference is used to purify the salt.

Salt from the rocks

Here are the stages that are used to purify rock salt.

A

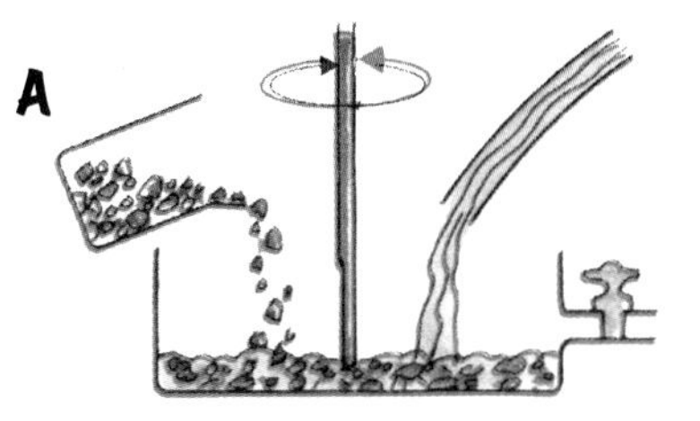

Stage 1 The rock salt is mixed with water.

The salt dissolves but the sand and clay are insoluble

B

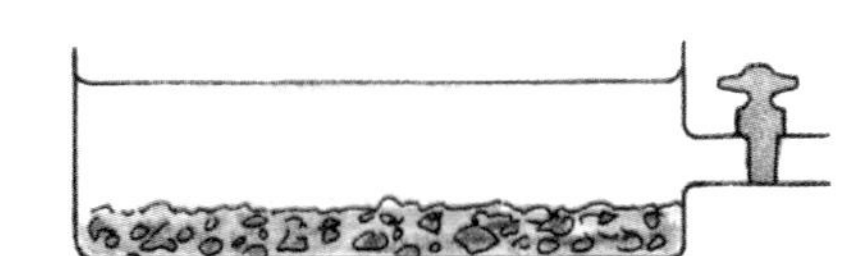

Stage 2 The mixture is allowed to stand.

C

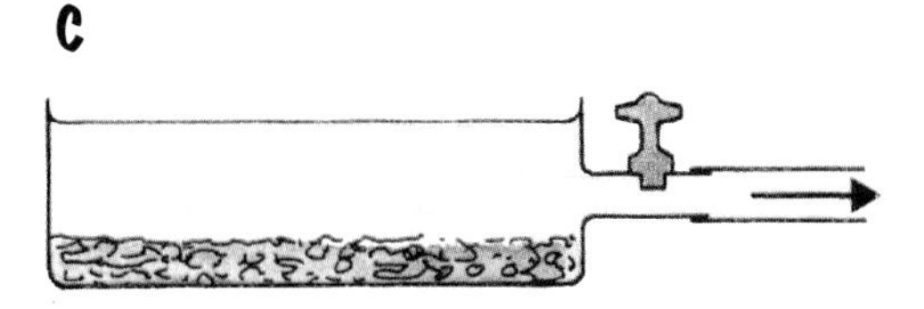

Stage 3 The liquid is run off.

D

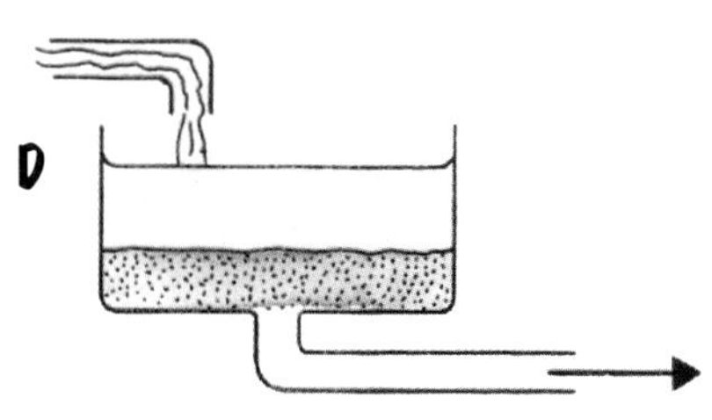

Stage 4 The liquid is filtered through fine sand.

E

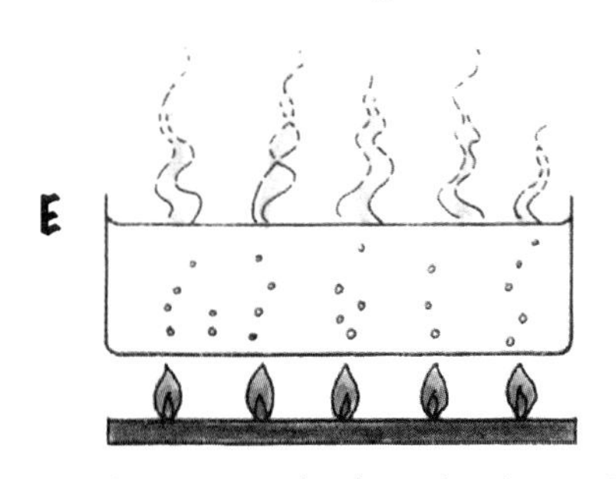

Stage 5 The liquid is heated.

1 Copy this diagram onto a full page and label the stages.

2 Under the labels, describe what is happening and how this helps to purify the salt. (The first one has been done for you.)

What do you know?

1 Copy and complete the following sentences. Use the words below to fill the gaps.

evaporated dissolve soluble
sink filtered solution

Salt is ________ in water, but mud and sand are not. Rock salt is mixed with water to ________ the salt. Most of the mud and sand ________ to the bottom, so the salt ________ can be run off. It is then ________ to get rid of the rest of the mud. The clear salt solution is ________ to give crystals of salt.

2 Where does the water go when a rain-puddle dries up in the sun?

3 Central heating can make the air very dry, and this damages house plants such as ferns. One way to overcome this problem is to stand the fern above a shallow tray full of water. Explain why this keeps the fern healthy.

4 How could you get copper sulphate crystals back from a copper sulphate solution?

Key ideas

Water will slowly **evaporate** into the air.

Evaporation happens faster if the water is heated.

If a solution is allowed to evaporate, the solute is left behind.

6d Getting the water back

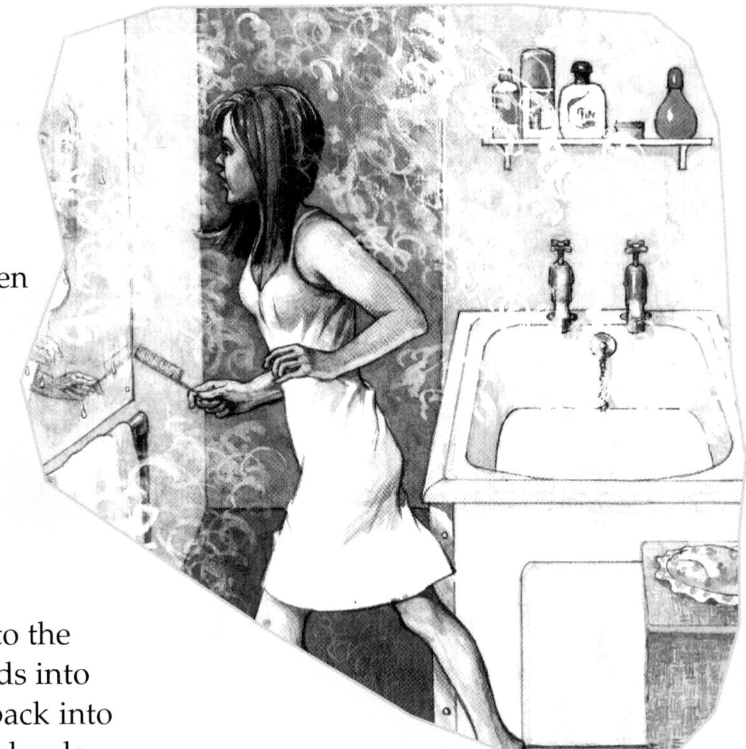

Evaporation is a good way to get the salt back from a solution. But what if it is the water you want? How could you get that back?

There is a clue in any steamy bathroom. Water vapour in the air is invisible, but bathrooms often become cloudy. What is going on? You can see the answer in the mirror . . .

Condensation

The water in a hot bath evaporates and goes into the air as water vapour. As the water vapour spreads into the room it cools slightly, and some of it turns back into tiny droplets of water. These hang in the air as clouds. The same thing happens when the warm water vapour hits a cold mirror or window. It cools down and turns back into liquid water. It **condenses**.

Make your own clouds

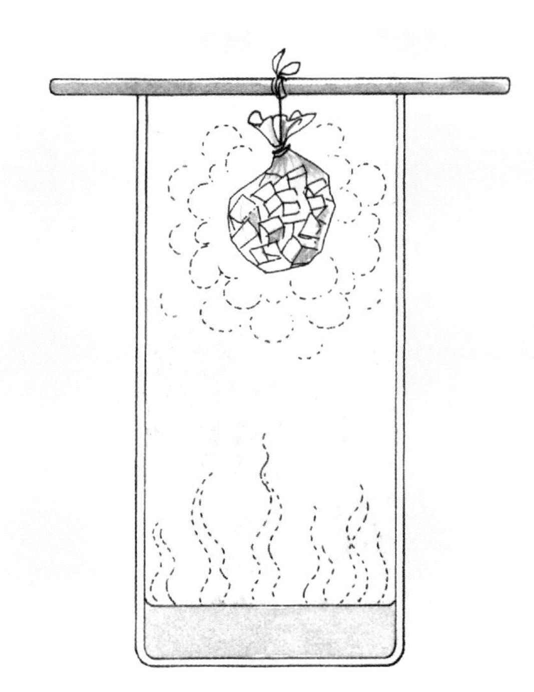

The clouds in the sky form in the same way as the clouds in your bathroom. Water vapour in the air cools and condenses as it rises up above the land or sea.

a You can make your own clouds in a tall jar. Just hang a bag of ice over some very hot water.

⚠ Take care
If you try this yourself, be very careful with hot water.

Collecting water

The warmer water is, the faster it evaporates. But if you heat the water until it bubbles and boils, you can quickly turn it all to gas (water vapour).

To get the water back, you collect the gas and cool it so that it condenses back to liquid water. This process of boiling and condensing water is called **distillation**. The pure water that is collected is called **distilled water**.

This diagram shows how to get pure water back from ink. The hot water vapour passes along a glass tube which is cooled by the air around it.

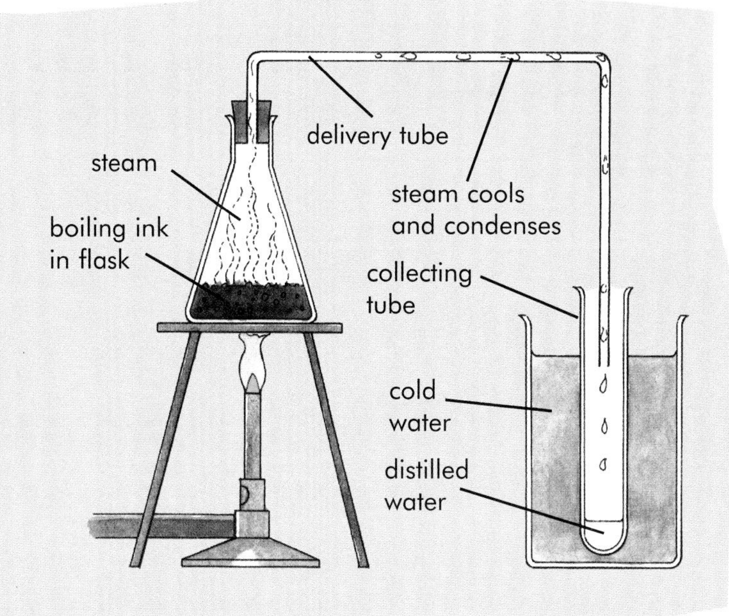

Why do you think the collecting tube stands in a beaker of cold water?

Surviving in the desert

Imagine you are stranded in the desert with no water and nothing but a clear plastic sheet and an empty jar. Never fear! Even in the desert, the soil contains some moisture. The heat from the sun can be used to collect this moisture.

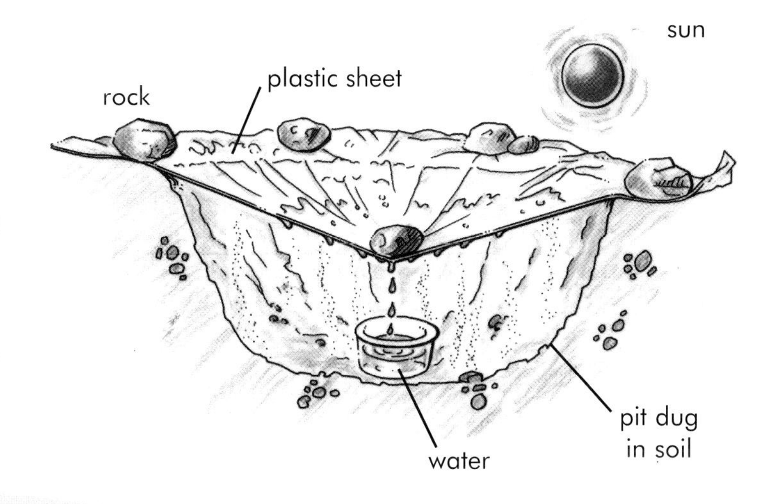

Try to explain how it works. You could even try it out in the garden.

What do you know?

1 Copy and complete the following sentences. Use words below to fill the gaps.

water vapour **liquid** **boil** **evaporation** **condenses**

Water can slowly turn into a gas at low temperatures by ________. If you keep heating, the water starts to ________. A lot of ________ ________ escapes into the air. If it is cooled it ________ and turns back to ________ water.

2 Wooden window frames in bathrooms need to be kept well painted or they quickly suffer from 'wet rot'. Why is this?

3 You are locked in a science laboratory all weekend with nothing to drink but ink. How do you survive?

Key ideas

Water vapour in the air can **condense** back to form liquid water if it is cooled.

Pure water can be collected from a solution by **distillation**.

6e What's in that mixture?

Have you ever spilt water onto your exercise book, or got it wet in the rain? If you use a felt tip pen, the ink dissolves in the water and makes a runny mess. Sometimes it also forms lots of different colours. Why does this happen?

Chromatography

If you put a drop of mixed ink onto filter paper and add water drop by drop, the water spreads out into the filter paper, carrying the ink with it. But the dyes in the ink do not move at the same speed. Some move out almost as fast as the water, while others lag behind. The result is that the dyes separate into a series of coloured rings. This process is called **chromatography**.

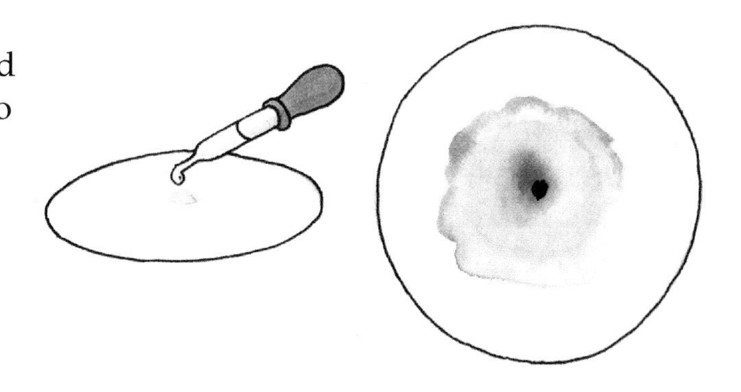

Chromatography at work

You can use chromatography to check the dyes used in sweets. Some yellow food dyes can be bad for some children, so they are taught to avoid yellow or orange sweets. But other coloured sweets could have the yellow dye in a mixture. How could you tell?

First dissolve some of the dye out of the sweet by standing it in water. Then evaporate some of the water to make the solution more concentrated.

Then compare the dyes from different sweets with the yellow dye. Put dots of the dyes on a strip of special chromatography paper, next to a dot of the yellow dye. Stand the paper in water. The water soaks up into the paper, carrying the dyes upwards.

Again, the dyes separate out into different colours. To check which sweets have the yellow colour, compare them with the yellow colour at one end.

Which sweets should be avoided?

Be a detective

Inspector Chroma thinks that this cheque has been forged. Somebody has changed six to sixty. The black ink looks the same, but perhaps the ink is different.

These cartoons show how Inspector Chroma investigated. Use them to plan your own experiment to compare the ink from different parts of a cheque.

What do you know?

1 Copy and complete the following sentences. Use the words below to fill the gaps.

speeds chromatography separates dyes

When water soaks up into paper, it carries any soluble ________ with it. The different dyes are carried along at different ________. This ________ out the different dyes. The process is called ________.

2 a Describe how you would find out whether the dyes used in Smarties were mixtures or not.

b If you had some samples of different red food dyes, how could you tell which of them was used in Smarties?

Key ideas

You can separate out mixtures of dyes using **chromatography**.

Chromatography can be used to identify the dyes in a mixture.

6 EXTRAS

6a Separating oil and water

When a cook roasts meat, a natural gravy forms. This has hot fat floating as an oily layer on top of watery stock. This gravy contains too much fat for a healthy diet. The liquids need to be separated, but how?

1 You can't just pour the fat off, because both liquids move together. You could spoon the fat off carefully, but that would take a long time. Invent a way of pouring off the stock without the fat.

6b More about solvents

Water is a very common solvent, but it is not the only one. There are many substances that are insoluble in water, but dissolve in other liquids.

If you get grease or oil on your clothes, you cannot wash it out because grease and oil are insoluble in water. But dry cleaners can remove the oil and grease. They use a different solvent called tetrachloroethane that can dissolve grease and oil. They are called dry cleaners because they don't use water.

Other useful solvents are white spirit, which dissolves gloss paint, and methylated spirits, which can be used to clean grease off things.

1 a Why is it no use to try to wash brushes with water if they have been used for gloss paint?

b What solvent should you use?

2 You have a mysterious patch of grease or oil on your T-shirt. How will you find out how to remove it?

Water does not always work!

6c Where does sugar come from?

Sugar cane is a bamboo-like plant with a sap that is rich in sugar. This diagram shows how sugar is produced from sugar cane.

1 The cane is crushed and mixed with water.

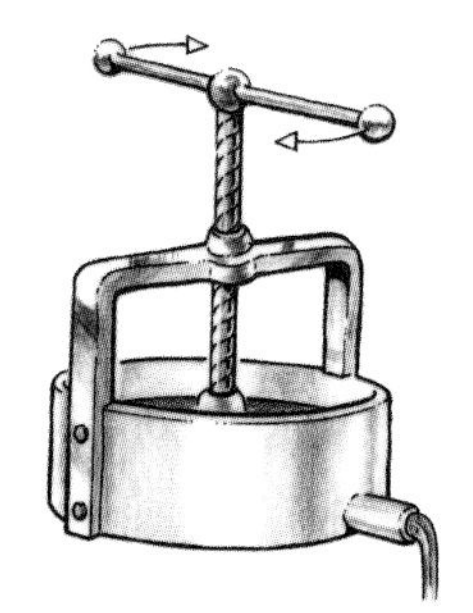

2 The liquid is squeezed from the pulp.

3 It is filtered.

4 The solution is boiled to give crystalline sugar.

1 Draw a simple flow diagram for the production of sugar from sugar cane. Explain what is happening at each step.

2 Sugar cane has a very woody stem. Once the sap has been removed, the dried stems are burnt as a fuel to help evaporate the sugar solution. What are the advantages of using them in this way?

6d The water cycle

Much of our weather is caused by water evaporating and condensing.

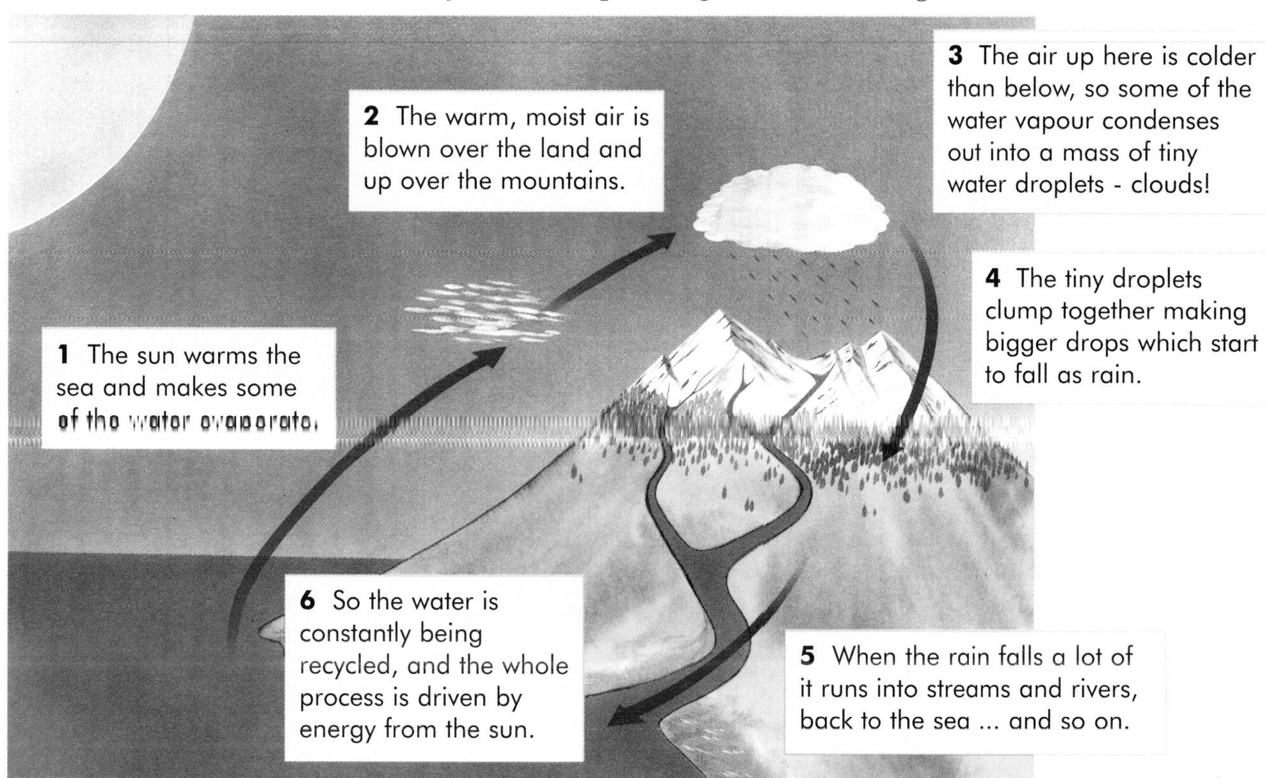

1 In Britain, warm, moist air often blows in from the Atlantic Ocean. Explain why this causes rain in the mountains (and elsewhere).

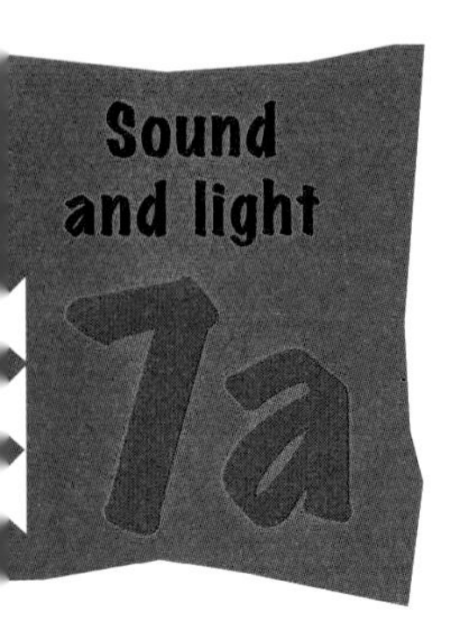

What causes noise?

An alarm clock like this makes a sound when the hammer hits the two gongs. A guitar makes a sound when its strings are plucked. But how do the gongs and the strings make the sound?

Making a noise

The gongs on the alarm clock and the strings on the guitar **vibrate**. All sounds are made by vibrations.

You also make sound yourself using vibrations. When you speak or shout or sing, the vocal cords in your throat vibrate. When the vibrations are bigger, the sound gets louder.

You can feel your vocal cords vibrating if you touch your throat while making a steady sound. It's easier to feel the vibrations if you sing a low note.

If you touch different parts of your head, you can feel vibrations when you make a sound. The bones of your skull are vibrating as well.

Noise annoys . . . and damages too

We call any sound we don't like a **noise**. Loud noises make big vibrations.

People who work in noisy places may suffer damage to their hearing. They need to wear ear protectors so that their hearing is not permanently damaged. If you listen to loud music a lot, you may damage your ears. Your hearing may never recover.

Ear, ear!

The vibrations of a sound travel through the air to your ears. They are collected by the outer parts of your ears. Some animals have big ears, so that they can hear faint sounds.

- Try cupping your hands round your ears. Can you hear more clearly?

Inside your ear, the sound vibrations make your eardrum vibrate. This pushes three little bones in the middle ear. The bones pass on the vibrations to the inner ear. Fine hairs in the inner ear vibrate and send a message to the brain.

Big vibrations from loud sounds can damage the bones of your middle ear and the fine hairs of your inner ear.

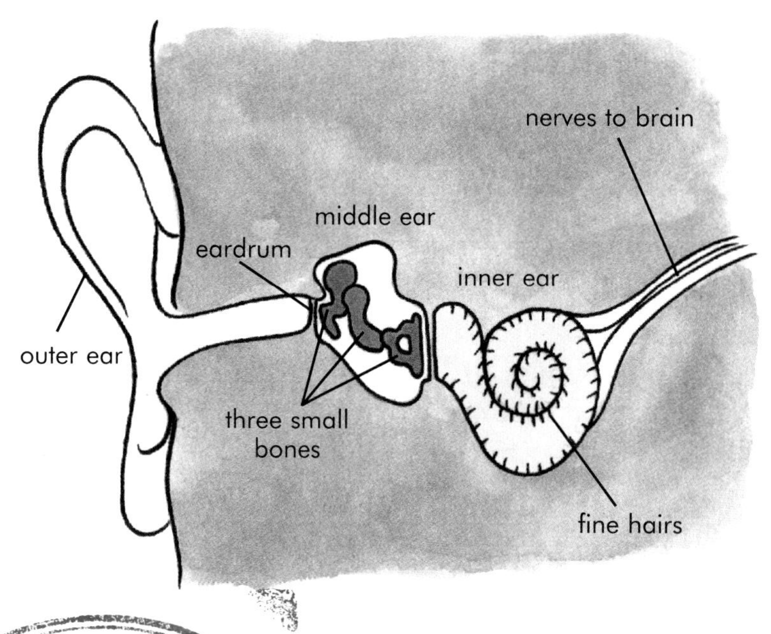

What do you know?

1 Copy and complete the following sentences. Use the words below to fill the gaps.

noise **eardrums** **ears**
vocal cords **sound**

When your _______ _______ vibrate, you make a _______. The vibrations travel through the air to your _______, where they make your _______ vibrate. Any sound which we don't like is called _______.

2 Look at the diagram of the ear. List the parts of the ear in the order in which they vibrate when a sound comes in, starting with the eardrum.

3 Your vocal cords vibrate when you make a noise. You can make noises in other ways, for example by hitting something so that it vibrates. Often the vibrations are too small to see. List five ways you can make a noise. For each one, say what vibrates.

Key ideas

Sounds are made when something **vibrates**. The vibrations are often too small to see.

You hear sounds because they make your eardrums vibrate.

Loud sounds can damage your hearing.

Less noise please!

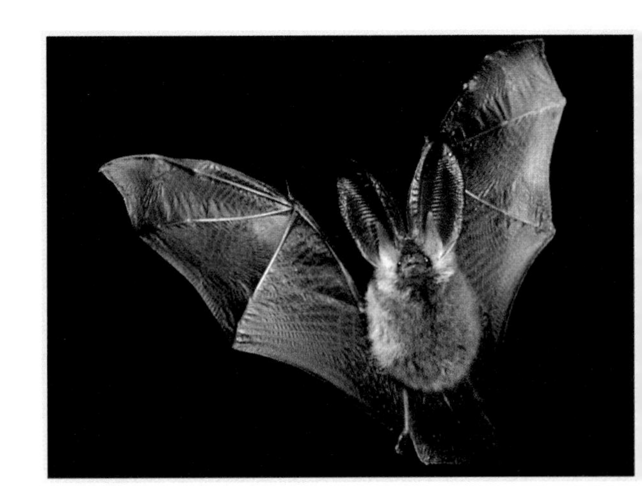

Bats do not have very good eyesight, but they do have excellent hearing. They live in a world of sound.

Instead of looking at their surroundings, bats listen to them. They make very high-pitched squeaks, too high for most humans to hear. Then they listen to the echoes of the squeaks coming back from their surroundings.

a Look at the photograph of the bat. What suggests that it has better hearing than you?

Hearing high and low

Bats make very high-pitched sounds. Some children can hear bats, but most adults can't. Dogs can also hear very high notes.

The diagram shows the **range** of hearing of different animals.

b **1** Which of these animals can hear the highest notes? Which can hear the lowest notes?

2 What happens to people's hearing as they get older?

bat

human adult

human child

cat

elephant

low notes — high notes

Reflecting and absorbing sound

Sometimes sounds come back to us from a high cliff or a building. This is what makes an **echo**. We say that the cliff or building **reflects** sounds back to us to make an echo.

Smooth, flat, hard surfaces reflect sounds well. They are good reflectors. That's why a big building with flat, hard walls and windows makes good echoes.

Soft, bumpy surfaces do not reflect sounds well. They **absorb** sounds.

c **1** Which materials in the following list are good reflectors of sounds?

2 Which are good at absorbing sounds?

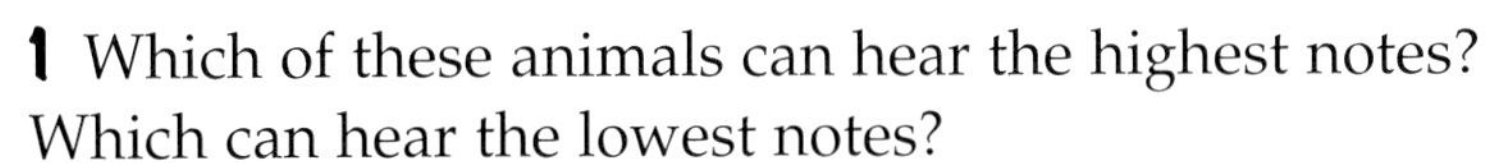

Bats can tell the difference between the clear echo from a hard flat surface and the muffled echo from a softer surface.

Cutting out noise

A bathroom can be a noisy place, especially if you sing in the bath. There are a lot of hard surfaces which reflect sounds. A bedroom is usually much quieter.

- What materials in the bedroom make it a quieter place? How could the bathroom be made quieter?

Improve your hearing

Have you ever wanted to listen to a conversation that is just too faint to hear? You can improve your hearing by making yourself some earflaps to collect up faint sounds.

- Design some earflaps to help you hear the faint ticking of a clock. Here are some points to think about:
  - How do animals' ears help them to hear well?
  - What would be a good material to use to make your earflaps?
  - What shape should they be?
  - How would you test your earflaps?
  - How would you compare them with other people's?

What do you know?

1 Draw a diagram to show how someone can hear an echo when they shout near a high wall. Draw a line to show how the sound travels from the mouth to the wall and back to the ears.

2 a What sort of surface reflects sound well? Choose words from the list below to describe surfaces that are good reflectors of sounds.

b Choose other words to describe good absorbers of sounds.

Record your ideas in two lists. (You don't have to use all the words.)

soft	**warm**	**smooth**	**bumpy**	**flat**
black	**cold**	**rough**	**hard**	**white**

3 Why do you think owls need to have very good hearing?

Key Ideas

Some people can hear higher and lower notes than other people. Young people usually have a wider **range** of hearing than older people.

Hard, flat surfaces are good **reflectors** of sounds. Soft, rough surfaces are good **absorbers** of sounds.

Travelling light

During the day, we can often see the Sun shining. In the photograph, you can see the Sun's rays shining down on to the Earth, lighting everything up.

a How can you tell from the photograph that the light from the Sun travels in straight lines? Where else have you seen light travelling in straight lines like this?

Night and day

At night, our side of the Earth is facing away from the Sun and we can no longer see its rays. Everything is dark. The Sun's rays cannot bend round and light up the back of the Earth.

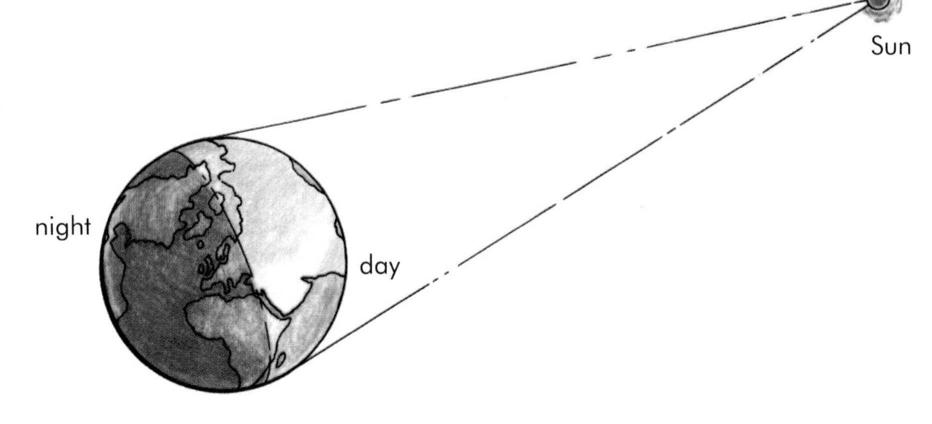

Shadow play

People on stage often make very clear **shadows**. The light comes from a spotlight. Its rays shine down on the singer to light her up. There is a dark shadow on the stage where the rays do not reach.

Because light travels in straight rays, we can draw a straight line to show where the singer's shadow will be.

Light and shade

You can make yourself look very spooky using the light from a torch. The light rays shine upwards from the torch. Some parts of your face are lit up, and others are in shadow.

- Try to work out how the strange shadows on your face are made.

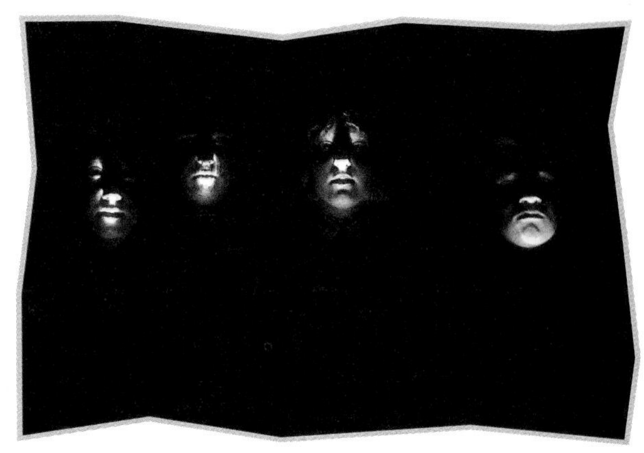

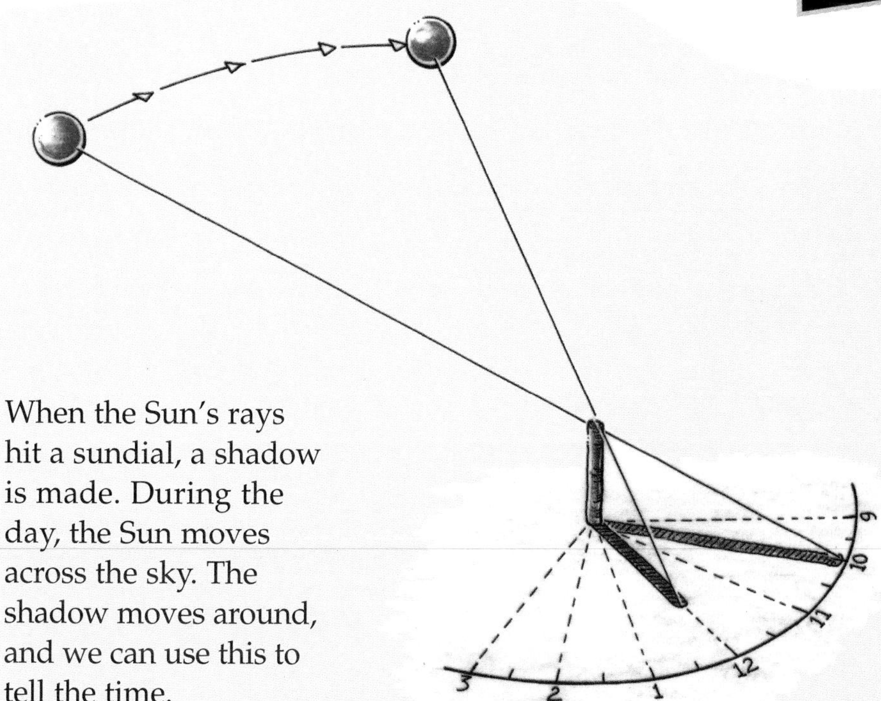

When the Sun's rays hit a sundial, a shadow is made. During the day, the Sun moves across the sky. The shadow moves around, and we can use this to tell the time.

What do you know?

1 Copy and complete the following sentences. Use the words below to fill the gaps.

shadows	**rays**	**lines**

Light travels in straight ________ called ________. When things block out the light, they make ________.

2 Some primary schools have a sundial marked on the playground. One pupil stands at the centre, and his or her shadow falls on the ground. The ground is marked to show the time. Write some instructions or draw a picture so that the pupils at a primary school could set up a sundial like this.

Key ideas

Light travels in straight lines.

When the path of light is blocked, a **shadow** is formed.

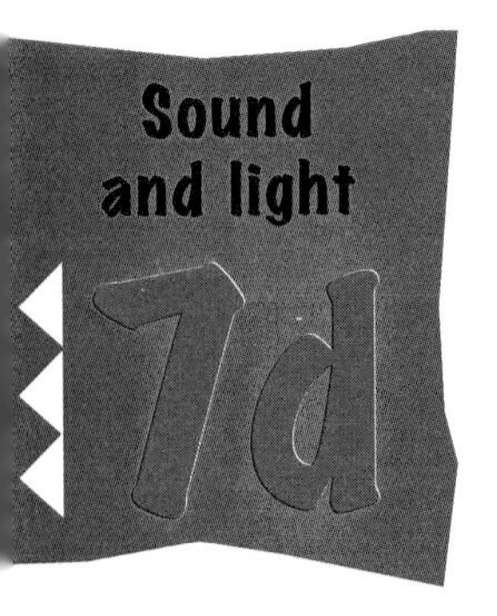

Long time coming

Have you ever noticed an interesting effect when watching a game like rounders or cricket? You see the bat hit the ball just before you hear it.

a Can you work out a reason for this?

Fast and faster

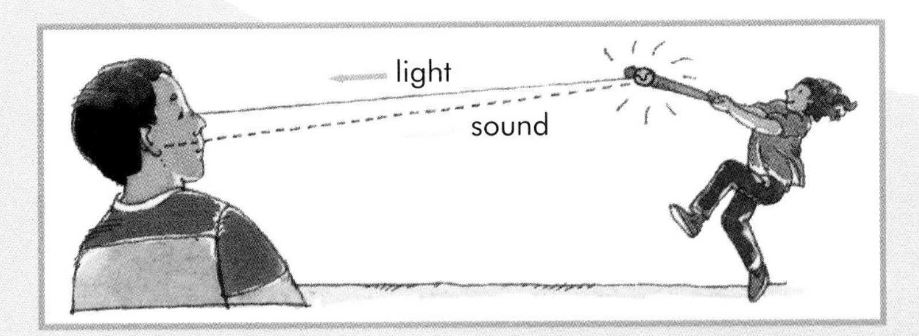

Sound and light both travel fast. The sound and sight of the bat hitting the ball reach you long before the ball does!

But light travels a lot faster than sound. In a thunderstorm, you see the lightning flash before you hear the thunder. The thunder is the sound made by the lightning flash, so they both happen at the same time. You can tell how close the thunderstorm is by how long it takes to hear the thunder after you have seen the lightning. Divide the number of seconds by three to find the distance in kilometres.

Light travel

The light from the Sun has to travel a long way to get to the Earth. It travels about 150 million kilometres! It takes the light about 500 seconds to get here. If the Sun stops shining, we won't know about it for 8 minutes!

The table shows how long it takes for light to travel some large distances.

Distance travelled	Time taken
around the equator	$\frac{1}{8}$ of a second
from the Moon to Earth	1.3 seconds
from the Sun to Earth	8 minutes
from distant stars	millions of years

Light travels very fast indeed. If you went in a spaceship to the Moon, it would take you several days. Light can make the journey in just over a second.

Astronomers study the light coming from distant stars. The stars that are the source of the light may be at the other side of the Universe. Their light has been travelling through space for billions of years. Astronomers can use it to find out what the Universe was like a long time ago.

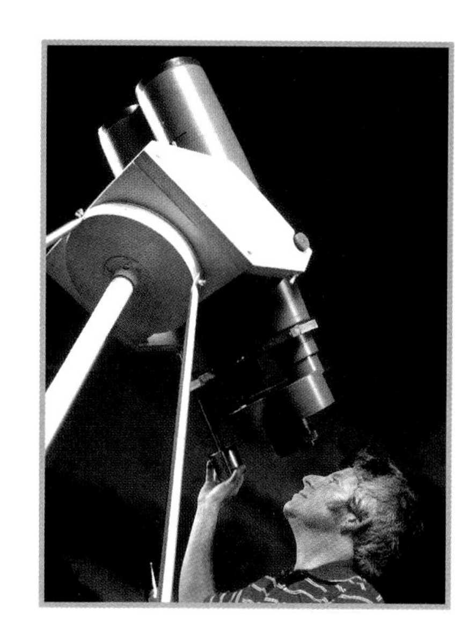

Long distance information

People have used light to send messages in many different ways. You can use a torch to flash a message in the dark. (That's why a torch is sometimes called a flashlight.)

Nowadays, many telephone messages are sent along glass fibres instead of wires. A tiny laser sends flashes of light through the glass. The light can travel from London to Edinburgh, over 600 kilometres, in less than a hundredth of a second.

What do you know?

1 Here are the contestants in an unusual race:

a tortoise   a sound   an Olympic sprinter
a ray of light   a racing car

Put them in order from fastest to slowest.

2 Look at the table on the opposite page.

a Which is closer to the Earth, the Moon or the Sun?

b What is the furthest object you can see during the day?

c What is the furthest object you can see at night?

d When can you see objects that are furthest away, at night or during the day?

3 In a stone quarry, they often use dynamite to blow up the solid rock.

a Explain why you would expect to see the explosion before hearing its bang.

b If the bang came six seconds after you saw the flash of the explosion, how far would you be from the quarry?

Key ideas

Light travels very fast, much faster than sound.

Bouncing light

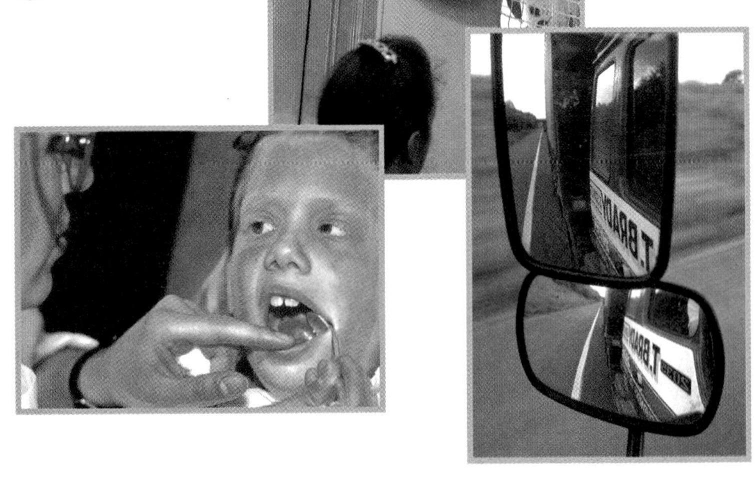

Some astronauts left a mirror on the Moon. Scientists on Earth shone a beam of light from a laser on to the mirror, and the light was reflected back to the Earth. It took the light less than three seconds to make the round trip of nearly a million kilometres.

a Who else uses mirrors at work? What do they use mirrors for?

Giving out light

We can see the Sun because it gives out light which shines into our eyes. The Sun is a **source** of light. Here are some other things that are sources of light.

Here are some other things that we can see. They are not sources of light. We need to shine light on these things to see them. The light is reflected back into our eyes.

Here are some ideas about how we see things.

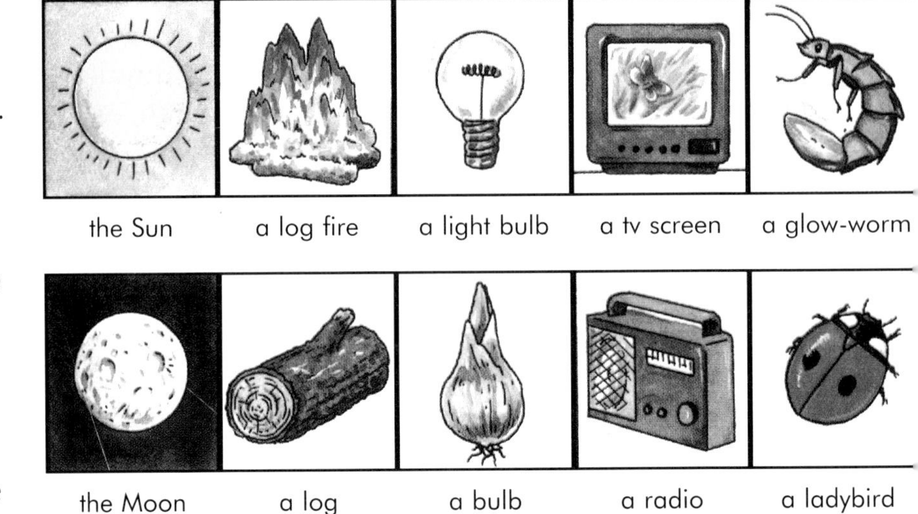

the Sun	a log fire	a light bulb	a tv screen	a glow-worm
the Moon	a log	a bulb	a radio	a ladybird

You just look at things to see them.

Light goes into my eye and onto the book, so that I can see it.

Light goes from the window onto the book, and then into my eye.

The last idea is the correct one. Light rays from the Sun shine on the book. They are scattered, and some go into your eye.

Rough and smooth

A rough surface **scatters** light rays, so that they go off in all directions. A smooth surface does not scatter the light. That is why a mirror is good at reflecting light.

Mirrors are very smooth. They reflect the light well. A smooth surface does not scatter the light. This means that the light comes back again in a straight line and we see . . . ourselves!

These two pictures show what happens when rays of light are shone on to some aluminium foil.

A

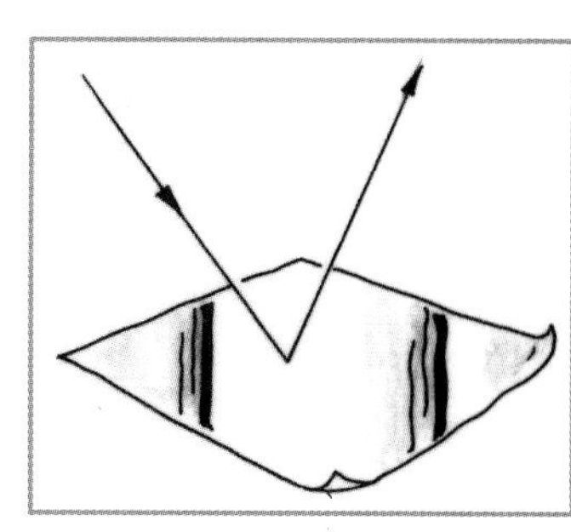

B

1 Which picture shows the rougher side?
2 Which shows the smooth side?
3 Which side will reflect the light better?

Moonlight

The Moon is cold and rocky. It does not give out its own light. We see it because it reflects the Sun's light. Some people think the Moon only comes out at night. The photo shows how you might have seen the Moon during the day.

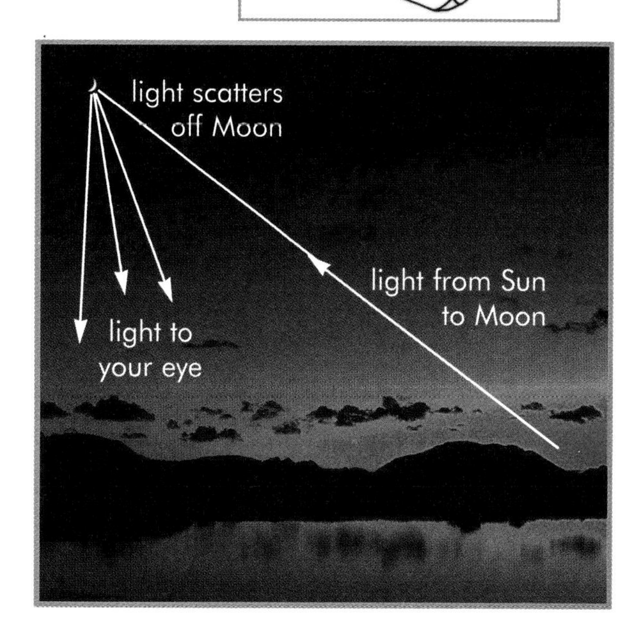

What do you know?

1 Copy and complete the following table. Use the words below to fill the gaps.

scattering **mirror** **pencil** **firework**

	a good reflector of light
	a bright source of light
	not a source of light
	happens when light rays fall on something rough

2 Draw a diagram showing what happens to the path of a ray of light when you see an object.

3 a How could you use two mirrors to see the back of your head? Draw a diagram to show how the two mirrors must be arranged.

b Draw a line to show how a ray of light scattered off the back of your head reflects off one mirror, off the other, and into your eye.

Key ideas

Light reflects well from smooth surfaces. It is **scattered** by rough surfaces.

A **source** of light gives out its own light. We can see other things because light scatters off them into our eyes.

EXTRAS

7a Better hearing

People who have poor hearing may use a hearing aid. A hearing trumpet collects more of the sound vibrations from the air. It is like having a bigger ear.

You may not notice if someone is using a tiny modern hearing aid. A microphone collects the sound, and an electronic circuit makes the vibrations bigger. A loudspeaker sends the bigger vibrations straight to the eardrum.

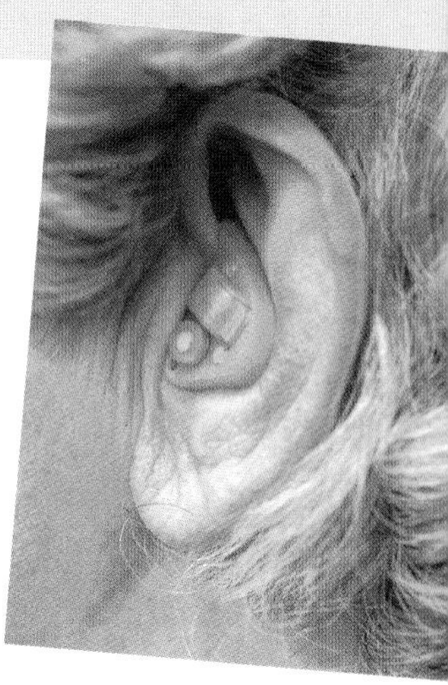

1 What can you say about the movement of the three bones in the middle ear when a person with poor hearing uses a hearing aid?

2 How could someone with a hearing trumpet use it to hear the voice of someone who is speaking behind them?

3 When you hear yourself speak, some of the vibrations travel through the air from your mouth to your ears. How else do vibrations reach your ears?

4 You may have noticed that your voice sounds different if you hear a tape-recording of yourself. Why do you think this is?

7b Sound direction

We have two ears for a very important reason. They help us to detect where a sound is coming from. If someone on your left is talking, the sound of their voice is louder in your left ear than in your right ear. Also, the sound reaches your left ear first.

1 Use a finger to close one of your ears. Test whether it is harder to tell where a sound is coming from. (It's best to shut your eyes as well.)

2 Cupping your hands around the back of your ears helps you to hear faint sounds. If you cup your hands round your ears the other way, from the front, you can catch sounds coming from behind you. Shut your eyes and try it. Where do sounds seem to be coming from?

3 You may have noticed that the sounds appear to be coming from in front of you when you cup your hands around the front of your ears. Draw a diagram to show how the sound reflects off your hands into your ears to explain why this happens.

7c Shadows large and small

This picture shows how a shadow puppet theatre works. The shadow is bigger than the puppet.

1 What will happen to the size of the shadow if the puppet is moved closer to the screen?

2 How should the operator move the puppet to make the shadow grow bigger?

3 How else could the operator make the shadow grow bigger?

Try to draw diagrams to explain your answers.

One person, two shadows

When actors appear on the stage, they may have two or even three shadows. This can happen if they have more than one light shining on them, from different directions. Each light makes a shadow.

You can create stage shadows using a toy person for the actor, and two or three torches for the lights.

4 Draw a diagram to show how three shadows would be formed.

5 What happens to the shadow if you make the light shine down from higher up?

Source or scatterer?

Young children may find it hard to understand that some things (such as a light bulb) are sources of light, and others (such as a book) can only be seen because light scatters from them into our eyes.

1 You have to help a young child to divide up a group of objects into those that are sources of light, and those that are not. Plan experiments to help them to do this.

Here are the things they have to separate:

fluorescent plastic stars   **a stone**
a glow-worm   **a torch bulb**
a Walkman   **a computer screen**

8a More about water

Water is very important. Two-thirds of the Earth is covered with it. You are about 70% water and would die very quickly without it. So how much do you know about water?

Water comes in three forms: **solid** (ice), **liquid** (water) and **gas** (steam or water vapour).

a Why is it like this? Think about where you find the three forms. Is there any pattern?

The missing link

You find ice where it is very cold. You find liquid water where it is not too cold and not too hot. You find steam where it is hot.

You can see that the missing link is **temperature**.

If you make water very cold, it freezes and turns to ice. If you warm it up again, it melts and turns back to water.

If you get water hot, it starts to evaporate. If you get it very hot, it boils and turns to steam. If the steam cools down again, it condenses back to water.

Measuring temperature

You can measure how hot or cold something is using a **thermometer**.

The temperature is usually measured in degrees Celsius (°C). On a very hot summer's day in England, the temperature might reach 30 °C. On a very cold winter's night it might fall to −5 °C.

The doctor sometimes takes your temperature if you feel ill. Your body temperature is 37 °C when you are healthy, but it might go up to 38 or 39 °C if you have the 'flu.

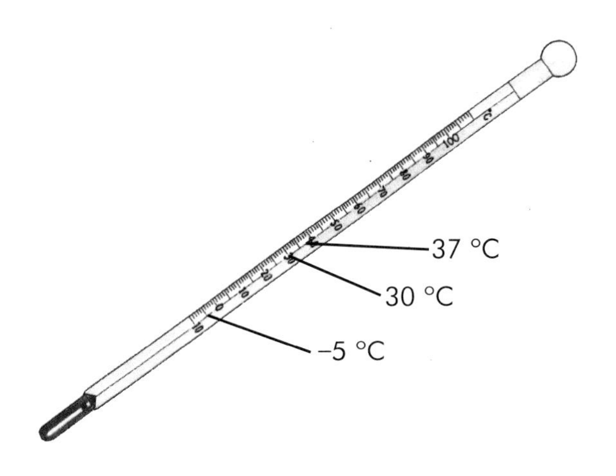

The melting point and boiling point of water

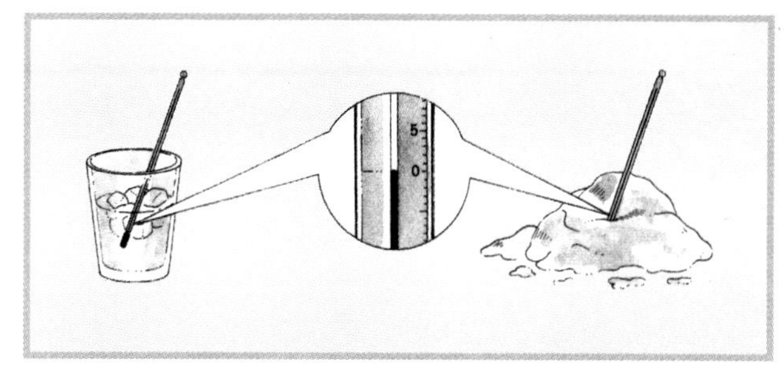

Ice always melts at the same temperature. Wherever you find melting ice, the temperature will be 0 °C. It works the other way around, too. If you cool water down, it will only start to freeze at 0 °C. This is called the **melting point** of water.

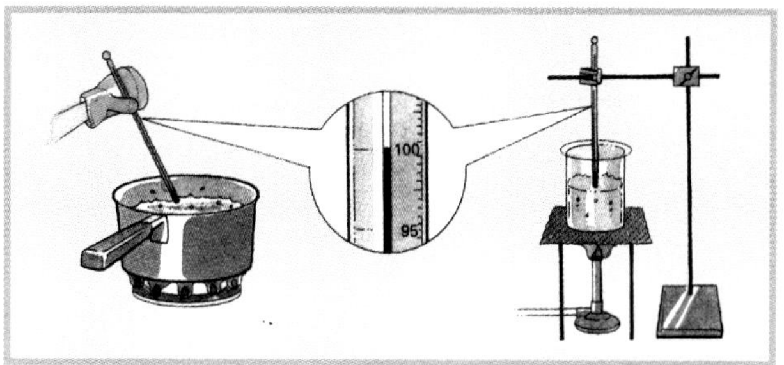

If you put a thermometer into boiling water, you will find that the temperature is 100 °C. This is the **boiling point** of water. Hot steam will condense back to water if it cools down to 100 °C.

Water can change into solid and then back into liquid, or from liquid to gas and then back into liquid. So these changes are called **reversible**.

Heating up the ice

Some students did an experiment to see how the temperature of a block of ice changed as it was heated up in an oven.

Here are the temperature readings in a graph.

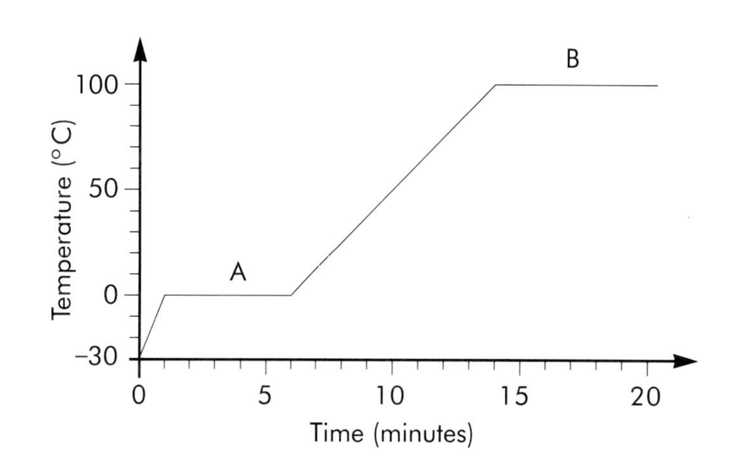

b **1** What is happening at the times marked A and B?

2 Explain how the line on the graph shows what is happening to the ice.

3 If you opened the oven door at the end of this experiment, what would you see?

What do you know?

1 Draw a thermometer like the one opposite and mark on it the melting point of water and the boiling point of water.

2 a Why does it turn frosty on a cold winter's night?

b Why is the water in your body a liquid?

3 The surface temperature on the planet Venus is 500 °C. In what form would you expect to find water there?

4 In an experiment, a clear liquid boiled at 80 °C. How do you know that it was not water?

Key ideas

Water can be a **solid** (ice), a **liquid** or a **gas** (steam), depending on its temperature.

Ice melts at 0 °C (its **melting point**) and water boils at 100 °C (its **boiling point**).

These changes are **reversible**.

8b Changing state

Like water, most things can exist as solids, liquids or gases. But unlike water, you don't usually see all three.

You know iron as a solid, and you may have seen liquid iron like the iron in this picture. You will not have seen iron as a gas. You know air as a gas and may know that skin-divers carry liquid air in their tanks. But have you ever heard of solid air?

Why are different materials found in different states like this?

Melting points and boiling points

Ice melts at 0 °C and water boils at 100 °C. But other things melt and boil at different temperatures. Iron melts at 1535 °C, so it is solid at room temperature. But the oxygen in air boils at −183 °C, so it is a gas at room temperature.

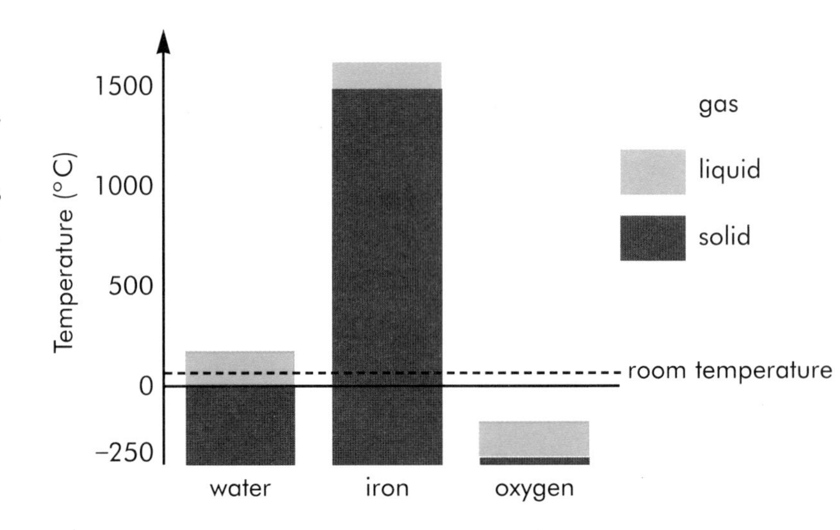

a Few things are liquid at room temperature. How many liquids can you think of that do not contain water?

Hot metal

Most metals have high melting points, but they can be melted in special furnaces. The molten metal turns back to solid when it cools. Metals can be shaped in this way. Molten metal is poured into a mould, where it sets to form a solid cast. Cast iron manhole covers are made in this way, and so are brass bells and bronze statues.

Casting is a useful technique, as scrap metal can be melted down and re-used, or recycled. But recycling is not a new idea. This giant statue in Le Puy in France was cast from recycled Russian cannons after the Crimean war, over 100 years ago.

Melting point and mixtures

Pure substances that are not mixed with anything else have fixed melting (and boiling) points. But if substances are mixed, these temperatures can vary. Water with salt in it does not start to freeze at 0 °C, it needs to get just a little colder. That is partly why rivers and ponds ice over before the salty sea during the winter. It is also why salt is put onto roads to stop them icing up in the winter.

The lower melting point of salt water helps keep roads clear in winter.

Which is the mixture?

Pure substances melt cleanly at one temperature. Mixtures soften and change over a wider range.

Which is the mixture, chocolate or candle wax? If you melt a small amount of each on a watch glass over a beaker of boiling water, this is what you see.

Which is the pure substance?

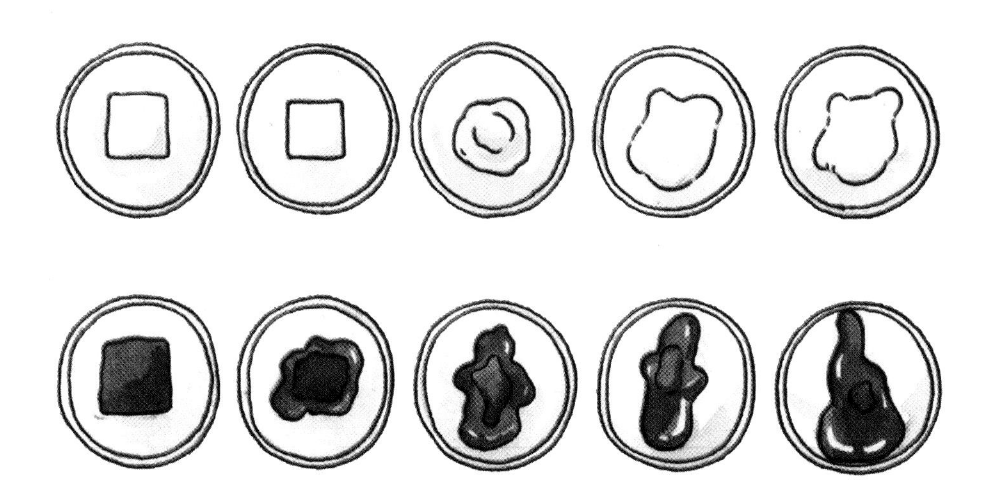

What do you know?

1 For each of the following, say whether it is a solid, liquid or gas at room temperature. (Room temperature is around 27 °C.)

	Melting point (°C)	Boiling point (°C)
a tin	232	2270
b bromine	−7	59
c fluorine	−220	−188
d butane	−138	−1
e propanone	−95	56
f benzoic acid	122	249

2 Explain how changes of state are used to shape metal.

3 Fat is 'solid oil' that sticks to plates and makes them difficult to wash up. It melts at about 40 °C. Why is it easier to do the washing up in hot water?

4 Vodka boils from 80 to 100° C. Is it pure alcohol? Explain your answer.

Key ideas

Different substances have different melting and boiling points.

Whether something is a solid, liquid or gas at room temperature depends on its melting and boiling points.

Changes of state such as melting and boiling are reversible.

Only pure substances have definite melting and boiling points.

Changing materials

8c Expansion and contraction

Have you ever grabbed a bottle of drink from the fridge to quench your thirst on a hot day, only to find that the cap was too tight to unscrew?

No problem! Just hold the cap under hot running water for a moment and it will unscrew easily.

What is happening here?

Expansion . . .

Most materials get a little bigger when they are heated, that is, they **expand**. When you run hot water over the bottle cap, the metal expands and loosens the fit, making it easier to unscrew.

. . . and contraction

If you now screw the hot cap on tight and put the bottle back in the fridge, you will have a problem the next time you want a drink. Expansion is reversible! As the cap cools down it shrinks back to its original size. It gets smaller again, that is, it **contracts**. This makes it grip the bottle tight again so it will be very hard to unscrew.

Expansion problems

Expansion can be a nuisance. For example, roads expand when they get hot. Some roads are built from concrete slabs. Each slab has to be separated by a gap filled with a rubbery substance (mastic) which squashes easily if the slabs expand.

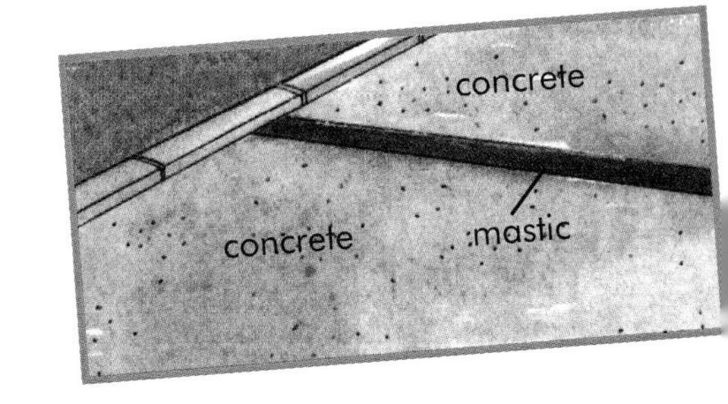

a What would happen in very hot weather if the slabs were pushed tight end to end without a gap?

Useful expansion

How can you fix a bicycle gear wheel tightly onto its axle? Make the hole for the axle just too small. Heat the gear wheel so that it expands, making the hole large enough to slip the axle into place. As the gear wheel cools and contracts, the hole gets smaller again, gripping the axle.

The same trick was used to fit the iron hoops onto wooden barrels.

Matched expansion

Use the information given here to answer the following questions.

If you took a 100 m length of each material and heated it by 10 °C it would expand by:

- concrete 11 mm
- steel 11mm
- Pyrex glass 3 mm
- brass 19 mm
- ordinary glass 9 mm

1 Concrete often has steel rods in it to make it stronger. Why is steel, not brass, used to reinforce concrete like this?

2 When you run boiling water into an ordinary glass bowl, the inside gets very hot, while the outside is still cold. The bowl breaks. Kitchen glassware is usually made of Pyrex rather than ordinary glass. Explain why.

How thermometers work

Liquids also expand when they are heated. You can see the effect if you fill a bottle with water and seal a clear straw or glass tube onto the top. When you heat the tube the water expands and goes up into the narrow tube.

You can measure how far up the tube the water goes at different temperatures, and mark a scale on it. It will be a water thermometer. Real thermometers work in a similar way.

What do you know?

1 Copy and complete the following sentences. Use the words below to fill the gaps.

buckle **gap** **cool** **expand**

When railway lines get hot, they ________.
There is a ________ between the lengths of rail to give room for this. If the rails were laid end to end, they would ________ when they get hot. When the rails ________ down again, they contract.

2 Expansion and contraction were used by old coopers (barrel makers) to fix the iron hoops on to their barrels. Explain how they did this.

3 Many thermometers use the liquid metal mercury, but others use coloured alcohol. Here are their melting and boiling points:

Liquid	Melting point (°C)	Boiling point (°C)
mercury	–39	357
alcohol	–117	79

Which type would you use:

a to find the boiling point of water

b to find the temperature at the North Pole in winter (which may be as low as –50 °C)?

Explain your answers.

Key ideas

Most materials **expand** when they are heated and **contract** when they are cooled.

The amount of expansion is small, but it can cause problems for large objects such as road slabs.

Liquids expand and contract in the same way.

8d Physical or chemical?

If you put a candle in an oven and heat it up, it melts. The wax has changed state, but it is still the same substance. When it cools down, you get the solid wax back. **Physical changes** like this are reversible.

If you light a candle, the wax melts at first, but then it seems to vanish. Where has the wax gone? It cannot be a physical change, as you cannot get the wax back. So what is happening?

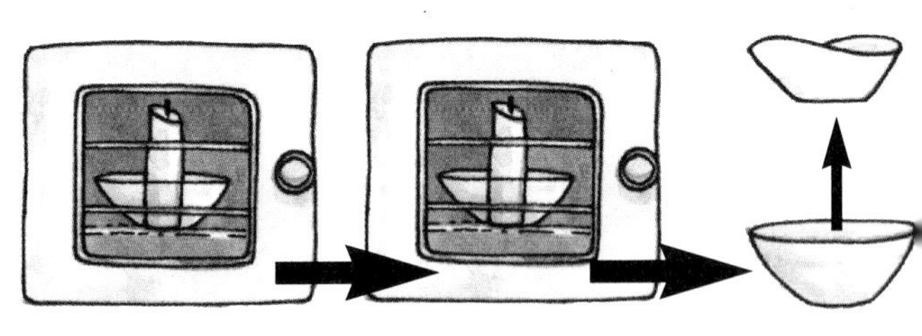

If you heat a candle, the wax just melts, so you can get it back.

The wax burns

When you light a candle, the wax burns. The heat from the flame makes the wax change into something different. This type of change is called a **chemical change**. Chemical changes are not easily reversible.

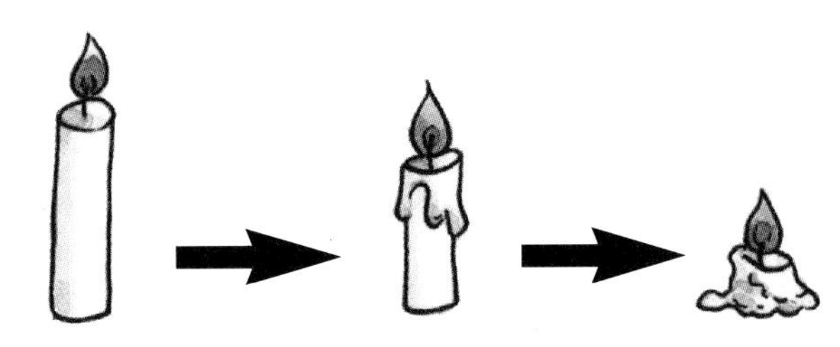

If you light a candle, the wax is lost, and you cannot get it back.

What are the new substances?

If you hold a beaker of cold water over a candle flame, you will see two new substances that are formed when a candle burns. The beaker gets covered in black soot. This is a form of carbon, like coal. You also see drops of water condensing on the cold glass. You might be surprised to find that water is formed when a candle burns!

A third substance is formed which you cannot see. It is the gas carbon dioxide.

water droplets

soot

How to spot a chemical change

This is not always easy! A change is probably a chemical change if:

- a completely new substance has formed, or
- the change cannot be reversed.

Toasting makes tasty new chemicals, but it makes carbon if you overdo it!

You can't get the wood back from a burnt match.

Giving off energy

Once the candle has been lit, it keeps burning, giving off heat and light. Chemical changes often give out energy, such as light, heat or even electricity.

Fireworks give off plenty of heat and light.

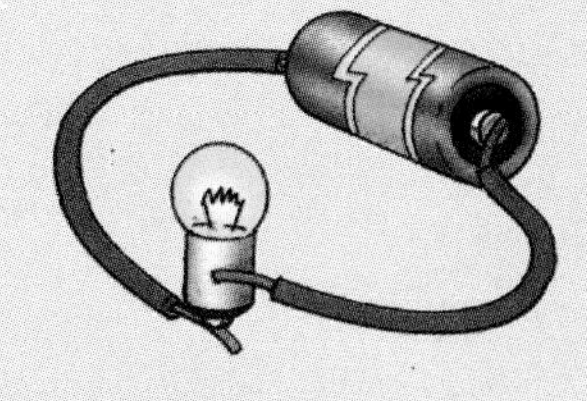

The chemical changes in a battery make electricity.

Physical or chemical changes?

a For each of these examples, say whether you think the change is physical or chemical. Give reasons for your answers.

What do you know?

1 Copy and complete the following sentences. Use the words below to fill the gaps.

chemical wax solid physical

When wax melts, it is a ________ change. The wax turns back to a ________ when it cools. Burning wax is a ________ change. New substances are formed and you cannot get the ________ back.

2 Are these physical or chemical changes? Give reasons for your answers.

a When you write with a felt-tip pen, the water evaporates from the ink, leaving the solid dye on the page.

b If you spill water on your page, the ink runs.

c If you try to dry it over an open fire, you set fire to the page.

3 Natural gas and candle wax are similar kinds of substances.

a If you heat a test tube with a yellow Bunsen burner flame, it will get covered in a black substance. What do you think this is? Explain your answer.

b Apart from this black stuff, what other evidence is there that burning gas is a chemical reaction?

Key ideas

Changes of state are **physical changes**.

Physical changes are reversible. **Chemical changes** are not reversible.

Burning is an example of a chemical change.

Chemical changes often give out heat, light or electricity.

8e More chemical changes

Where does all that energy come from when the Space Shuttle blasts off? It comes from a simple chemical reaction that is similar to a burning candle. The Space Shuttle is burning a simple **fuel** called hydrogen.

Burning fuels

When you burn a fuel, you get a lot of **energy**. Burning fuels is a very important chemical reaction. Your home would be cold and dark without it! The fuel reacts with the gas **oxygen**, and new substances are formed. You can write this reaction down in a kind of shorthand called a chemical equation.

fuel + oxygen ⟶ new chemicals + energy

When the Space Shuttle burns its hydrogen, the new chemical is water. This comes off as steam.

Where is the oxygen?

The Space Shuttle carries its own oxygen in special tanks. When you burn coal, the oxygen comes from the air. One-fifth of the air is oxygen.

Coal is a solid fuel which is mostly carbon. When carbon burns, it reacts with the oxygen in the air to form a new chemical and give out lots of energy. The new chemical is the gas carbon dioxide.

Burning other fuels

Other fuels we use are gas, oil and wood. These have more complicated particles than coal, made from carbon and hydrogen. But when they burn, the carbon still turns to carbon dioxide and the hydrogen turns to water. They give out lots of energy for heating, or for moving a car.

Not all chemical reactions are useful

Oxygen is also involved in some reactions that we could do without! One reaction causes great problems and costs a fortune to put right. This is rusting iron.

Iron and steel are the most commonly used of all metals. You see these metals everywhere in bridges, railways, machinery, girders and cars. But if you look closely you will also see the tell-tale brown stains of **rust**. Left unprotected, iron and steel will rust away completely over the years. Machinery seizes up, bridges crumble and cars fall to pieces.

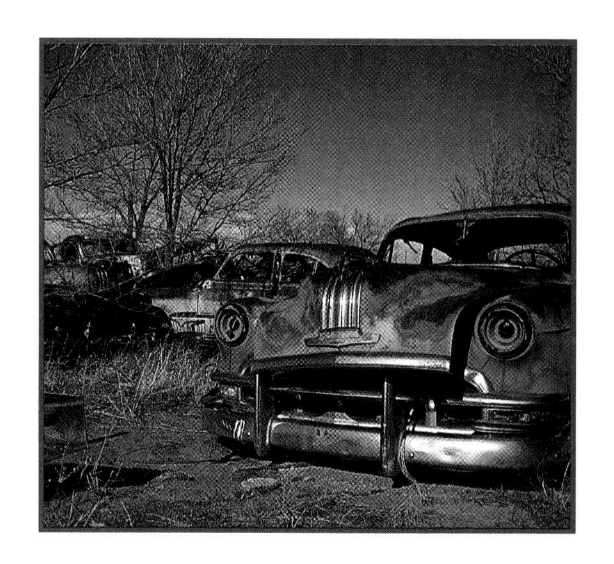

What makes iron rust?

Kim set up three tubes with an iron nail in each.

- The first had dry air (no water).
- The second was full of boiled water (no air).
- The third was half-full of water (air and water).

a What is needed to make iron rust?

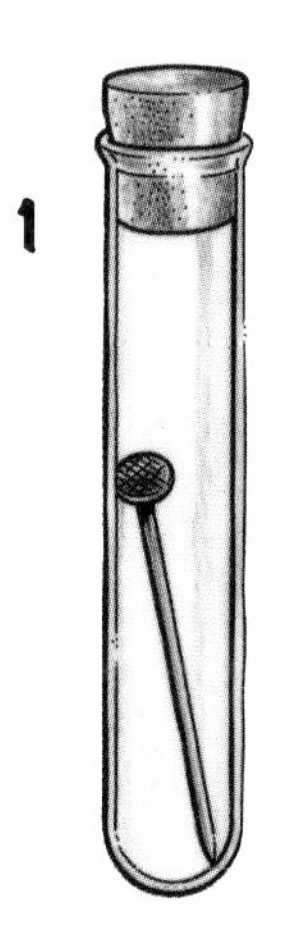

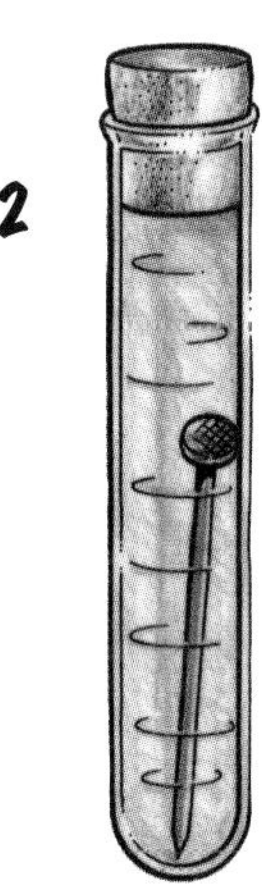

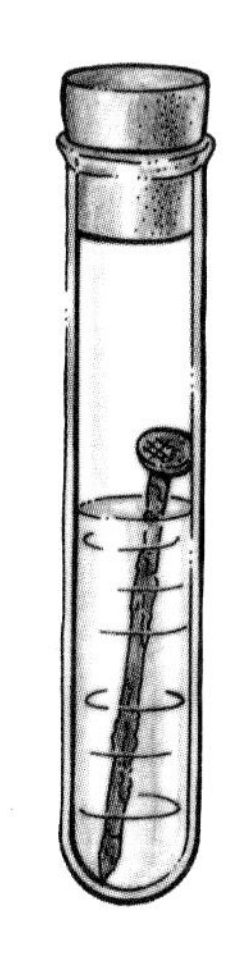

After one week this is what the tubes looked like.

What do you know?

1 Copy and complete the following sentences. Use the words below to fill the gaps.

chemical energy oxygen fuel

When a ________ burns, it reacts with the ______ in the air. Burning is a ________ reaction which gives out ________ as heat and light.

2 Wood contains hydrogen and carbon.
a What are the substances that form when wood burns in air?
b Write a word equation to show what happens.

3 Where would you expect cars to rust fastest, in the rainy tropics or the desert? Explain your answer.

Key ideas

Some chemical reactions are very useful.

Fuels burn in **oxygen** and give out **energy**. The oxygen usually comes from the air.

One-fifth of the air is oxygen.

Some chemical reactions are not useful. Iron turns to **rust** in the presence of air and water.

EXTRAS

8a Rain, snow and hail

Clouds form when tiny droplets of water condense out of the air, and these droplets collect to form raindrops.

Over cool countries it is well below freezing where the clouds form, so the water condenses out directly as tiny crystals of ice. Usually these melt as they fall, giving rain.

Where large clouds form, the air is usually moving upwards through them. Sometimes ice crystals fall through the cloud and melt into raindrops. These raindrops get blown back upwards, so that they refreeze into balls of ice, hailstones.

In some very large rainclouds, this can happen many times. Each time round, the hailstone collects a little more water, which then freezes to form a new layer of ice. This makes the hailstones grow, like mini snowballs.

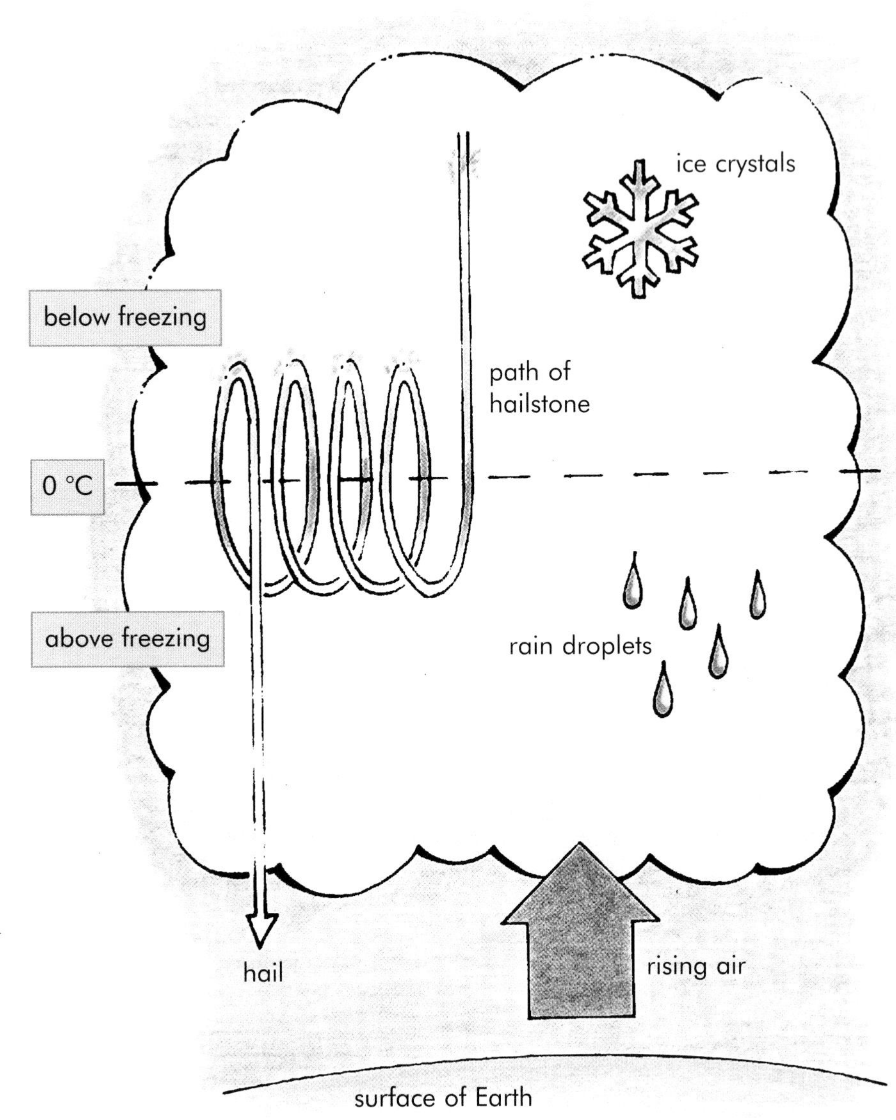

1 Copy the diagram above and label it to explain how hailstones form.

2 If you cut a large hailstone in half, you will see that it has rings like a tree trunk.
a What causes these rings?
b If a hailstone had five rings, what would that tell you?

3 Why do you not get hailstones on a freezing cold day?

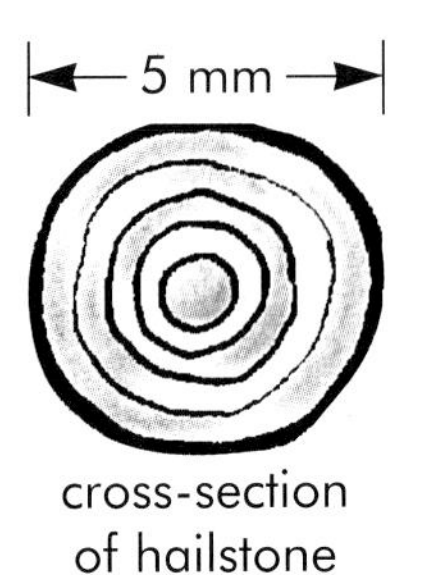

cross-section of hailstone

8c Tricks with expansion

A bimetallic strip has two different metals, such as brass and invar, welded together. Brass expands nearly 20 times as much as invar, so if the strip is heated, it bends. (The outside of the bend is longer than the inside, like the lanes on a running track.)

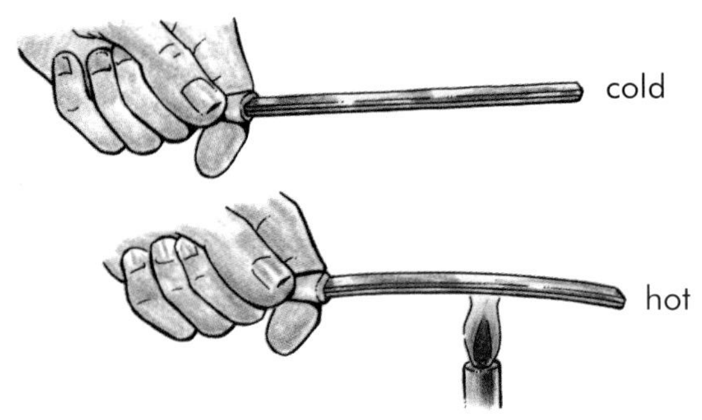

A bimetallic strip can be used as a switch in an electrical circuit, to turn on a fire alarm, for example.

1 Some heaters have bimetallic strips to switch the electricity off if it gets too hot. How could you change the circuit shown below so that it switches the alarm off when it gets hot, instead of switching it on?

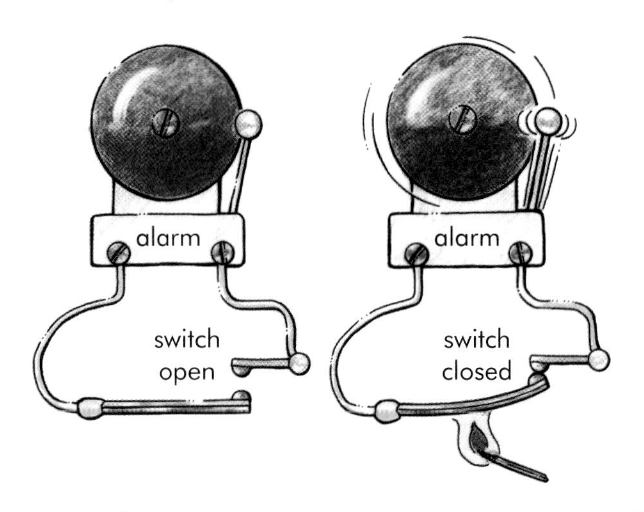

8e Chemical change and food

The food you eat contains complex substances that must be broken down into simpler ones before your body can use them. Your digestive system does this, but some foods that come from plants are quite hard to break down (see Unit 5). Cows overcome the problem by having three stomachs. Humans cook their food instead!

Cooking involves heating the food. Heating breaks large particles into smaller ones. The starch in bread can be broken down into sugar, for example. Cooking also softens up the food, breaking down its structure to help your digestive system.

1 Plan an experiment to see what effect boiling has on carrots or potatoes. You could cut equal sized pieces and boil them in water for different times. How could you measure how soft they had become?

2 Why does toast taste sweeter than bread?

3 Rabbits don't cook their grass. They have a different way of overcoming the problem of eating food that is hard to digest. See if you can find out what it is.

9a More of the same

All living things eventually die. So while they are fit and strong, all living organisms need to make more of themselves (**reproduce**). If plants and animals didn't make more plants and animals there would soon be no living things at all.

Some living organisms make more of themselves by simply splitting in two. Many living things use a more complex method called **sexual reproduction**.

In sexual reproduction, a special cell from the female joins with a special cell from the male to make a new and different cell. This cell will grow into a new plant or animal.

Special cells for a special job

The special cells that take part in sexual reproduction have to join together and form a new life. They are called the **sex cells**. When they join together it is called **fertilisation**.

Female sex cells are always bigger than male sex cells, but there are usually many more male sex cells than female ones. The male cells have to travel to the female cells. They may be carried by air or by insects, or they may travel under their own steam.

In plants, the big female sex cell is called an **ovule** or **egg**. It stays in the parent plant. The small male sex cell is called **pollen**. It is carried from one flower to another by the air or by animals such as insects.

In humans and other animals, the big female sex cell is called an **ovum** or **egg**. The small male sex cells are called **sperm**. They leave the body of the male and swim towards the egg by lashing their long tails.

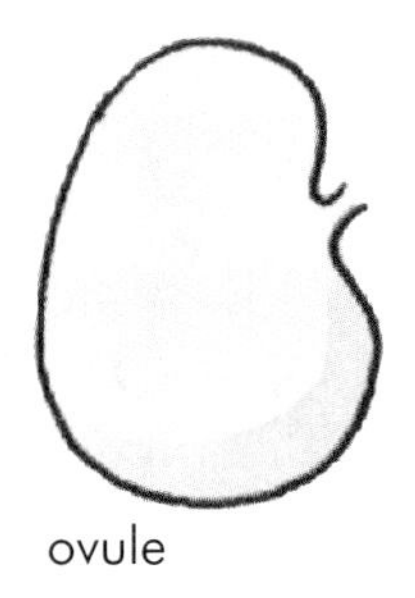

Sex cells in plants

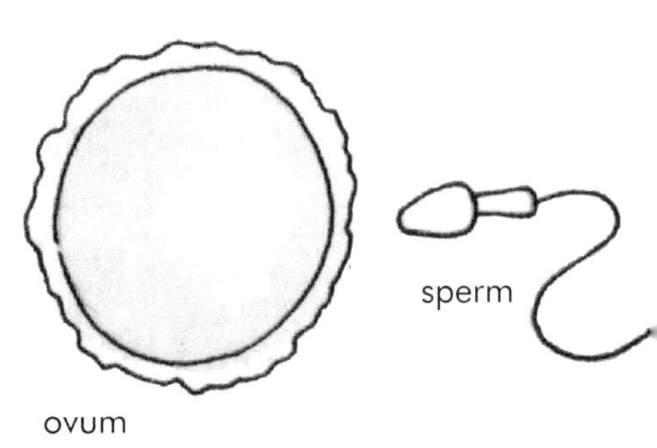

Sex cells in animals

Two's company

How do the female and male cells find each other? This is a problem that different organisms solve in different ways.

I attract bees. They carry my pollen to another flower. They also bring me pollen.

My pollen gets blown away in the wind. My sticky female parts catch another grass's pollen.

My tail convinces the females that I'm the best male around. Then we mate. I put my sperm inside the female so that it has a good chance of getting to the egg.

We just pump our eggs and sperm into the sea and hope for the best!

What do you know?

1 Copy and complete the following sentences. Use the words below to fill the gaps.

sexual **reproduce** **sex** **plant**

All living things need to ________. Many use ________ reproduction. This means two special cells called ________ cells join together to form a new animal or ________.

2 Link the descriptions below to the names.

Descriptions	Names
the male sex cells in animals	ovum
the female sex cell in animals	pollen
the male sex cells in plants	sperm
the female sex cell in plants	ovule

3 Describe two methods plants use to get their pollen to the ovule, and two methods animals use to get their sperm to the ovum.

Key ideas

All living organisms need to **reproduce**.

Sexual reproduction involves special female and male sex cells.

The female sex cell in a plant is called an **ovule**. In an animal it is called the **ovum**.

The male sex cells in plants are called **pollen**. In animals they are called **sperm**.

Fertilisation is the joining of a male and female sex cell.

9b Flowers

Flowers are part of our life. We use them as presents to say 'thank you' or 'I love you' and to make our homes, gardens and parks nicer places to be. But flowers are not just attractive to look at. They play an important part in the reproduction of plants.

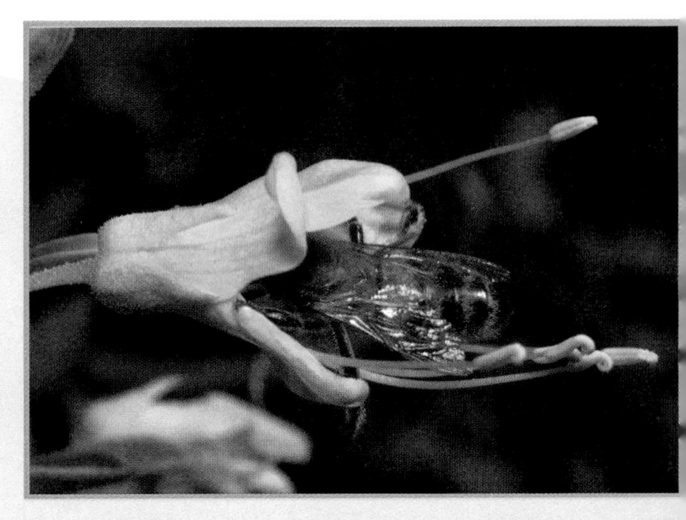

The secrets of a flower

Flowers like these use insects to transfer pollen from one flower to another. Pollen is made in the **stamen**. The insects are attracted by the bright petals and go right into the flower to get at its sweet nectar. Pollen rubs off the stamen onto their bodies.

In the next flower they visit, this pollen rubs off the insect onto the **stigma**. At the bottom of the stigma is the **ovary** which makes the ovules. When the pollen from one plant has landed safely on the stigma of another plant, **pollination** has taken place.

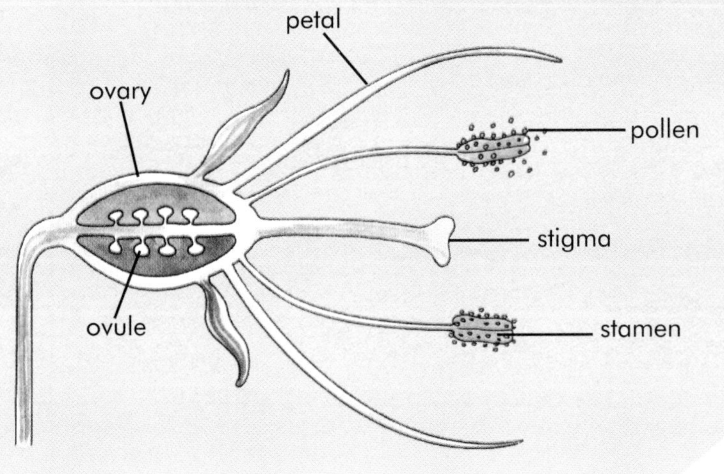

Getting things together

Not all plants attract insects. Plants without bright petals rely on the wind blowing their pollen about.

This is a drawing of a wind pollinated plant. Copy it and add labels showing the petals, stamen, stigma and ovary.

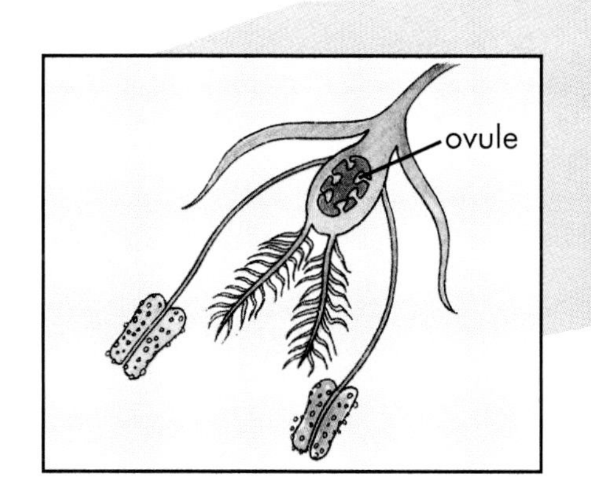

What happens next?

Once a plant is pollinated, the male sex cell moves to the ovary and joins with an ovule (egg). This is **fertilisation**. The new cell that is formed grows and divides lots of times to form a **seed**.

Inside the seed a tiny new plant called the **embryo plant** develops. The seed has a tough coat around it to protect the embryo plant. It also has a good store of food. This supports the tiny new plant until it grows out of the seed and can make food for itself by photosynthesis.

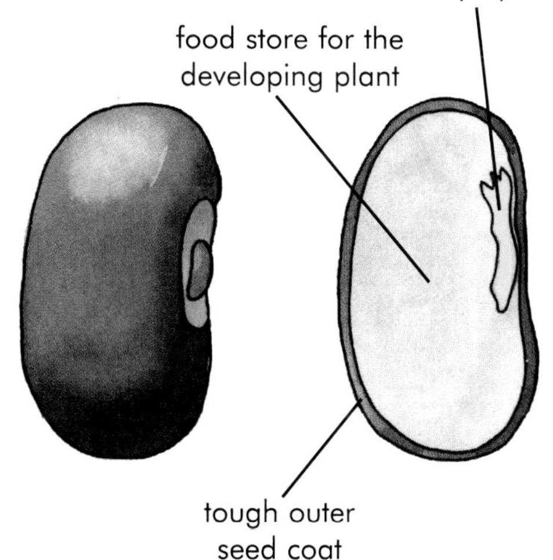

What do you know?

1 a I am brightly coloured to help attract insects to my plant. What am I?

b I make the female sex cells. What am I?

c I make the male sex cells for my plant. What am I?

d I am the male sex cells of this flower. What am I?

e Male sex cells from another flower are carried to me. What am I?

2 When pollen from one plant has been collected by the female parts of another plant, what has taken place?

3 One flower produces lots of small, smooth pollen grains. Another plant produces a small amount of big, spiky pollen grains. Work out which flower is insect pollinated and which is wind pollinated. What helped you to decide?

Key ideas

Flowers contain the sex organs of plants.

Pollen made in the **stamen** of one plant lands on the **stigma** of another. This is called **pollination**. The pollen **fertilises** the ovules in the **ovary**.

The fertilised ovules develop into **seeds** each containing a new **embryo plant**.

9c Scattering the seeds

As seeds grow, so does the ovary that contains them. The ovary forms a **fruit** which protects the developing seeds.

a Make a list of ten fruits you can think of. How many of them can you eat? Where are the seeds?

Fabulous fruit

Many of the fruits you thought of were probably the sort we eat. The part of a cherry we eat is the swollen ovary. The stone is the seeds.

Because the seeds form inside a fruit, they are protected from disease. But more importantly they can be carried away from the parent plant (**dispersed**) before they start to grow.

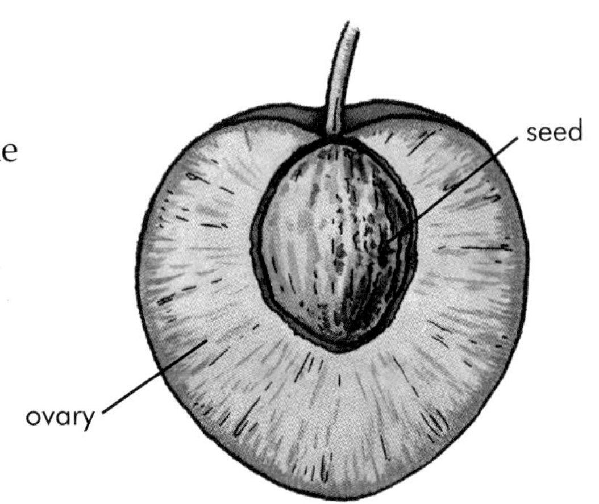

Why are seeds dispersed?

To grow well, plants need light and air, and water and minerals from the soil. Plants that have been growing for some time have big deep roots which can draw water from a large area of soil around the plant.

b **1** If seeds simply dropped to the floor beneath their parent plant, they would have problems getting some of the things they need. Which ones?

2 Why is it important for seeds to grow some distance away from bigger plants?

Carried away

Some fruits, like apples and blackberries, are sweet and fleshy. The fruits are eaten by birds and other animals. Some time later the seeds pass out of the animals, landing on the soil with their own supply of manure!

Sometimes the ovary becomes very hard, forming a nut. Animals carry the nuts away and hide them to eat later. The animals often forget where they have hidden the nuts, so later the seeds begin to grow.

If the ovary wall is sticky or hooked, the whole fruit gets stuck to an animal. Later, when the animal grooms itself, the fruit falls off and the seeds can grow.

Some fruits have wings or fluffy parachutes which help them to float away from the parent plant on the wind.

What do you know?

1 Copy and complete the following sentences. Use the words below to fill the gaps.

fruit **parent plant** **flower** **ovary** **dispersed**

Once fertilisation has taken place in plants, the ________ dies away. The seeds continue to develop inside the ________. The ovary itself also grows to form the ________. Once the seeds have formed, they are ________, which means they are carried away from the ________ ________.

2 Why do fruits form around seeds?

3 On a walk you find three different types of fruits that you've never seen before. One is very light and small with a fluffy parachute. Another has a very sticky coat, and the third has a very hard and woody outside. Work out how the seeds are dispersed for each fruit.

Key ideas

Seeds develop inside **fruits** which form from the ovaries of plants.

The fruits help the seeds to **disperse** using the wind or animals.

Seeds are dispersed away from parent plants to find their own light, water and minerals.

9d From boy to man

Young human beings look very similar, particularly when they have their clothes on. Parents often dress baby boys in blue and baby girls in pink just to avoid confusion! Few people would make the same mistake with adult humans.

a Look at the picture of young adults. What clues help you to decide which is male and which is female?

Why change at all?

All the special organs you need for making and growing babies are there at birth, but they need to develop to work properly. Your body changes from that of a child to that of an adult ready for reproduction at a stage in your life called **puberty**.

Male reproductive organs

Men have **testes** which make **sperm**, the male sex cells. The testes are held in a bag of skin called the **scrotum** outside the body to keep them cool.

The sperm travel up the sperm ducts from the testes and mix with a liquid produced by the glands. The sperm and liquid together are called **semen**.

This travels out of the man's body through a tube in the penis called the **urethra**. (This tube also carries urine away from the body.) Once a boy's body begins making sperm, it will carry on for the rest of his life.

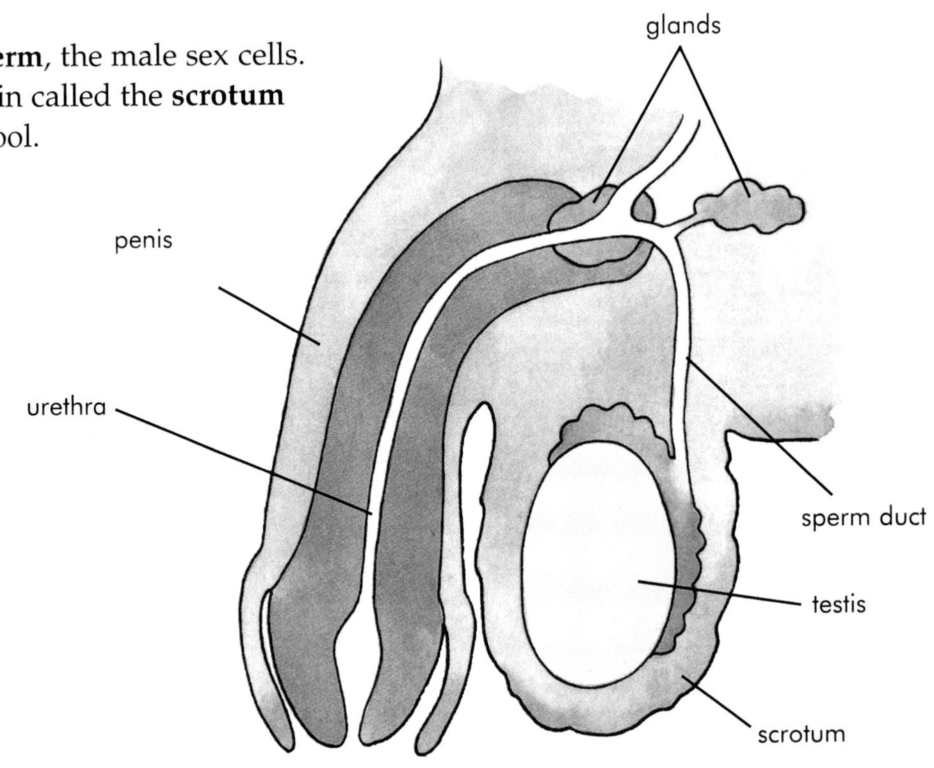

Boys become men

Puberty in boys usually begins between the ages of 11 and 16. It may happen very quickly or very slowly and is different for everyone.

During puberty **facial hair** (a beard and moustache) starts to grow. It can take several years before a boy needs to shave every day. **Body hair** also appears. Some men are naturally very hairy, while others are quite smooth.

The Adam's apple (**larynx**) in a boy's throat gets larger, and so his voice becomes deeper. This can happen gradually or very suddenly.

His **sex organs** (the **penis** and the **testes**) grow larger and the skin on them darkens. His testes start making sperm. **Pubic hair** grows around his sex organs and under his armpits.

His **body shape** changes slowly as the bones and muscles grow, so that his shoulders and chest are broader than his hips.

Boys also grow very quickly during puberty and get close to their adult height.

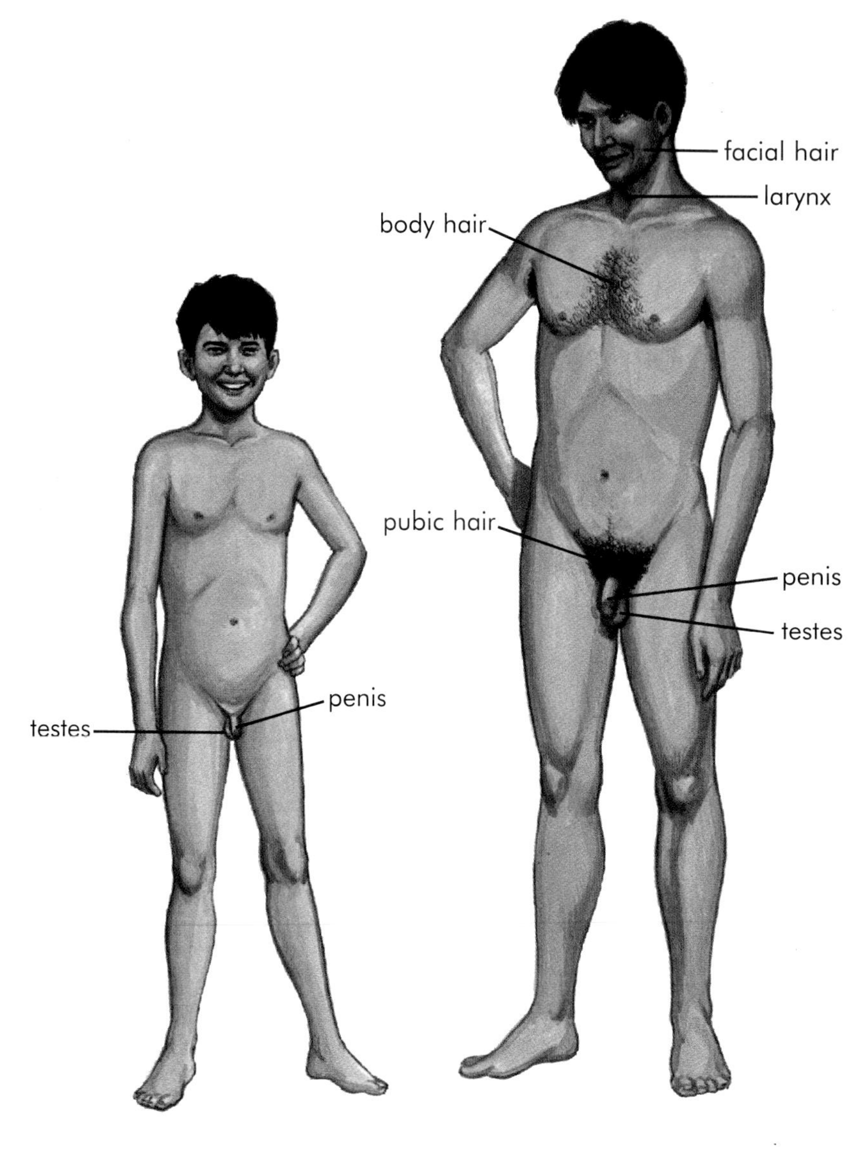

What do you know?

1 Make a list of all the changes that take place in a boy's body at puberty.

Key ideas

Puberty is the stage when your body changes from that of a child to that of a young adult capable of reproducing.

Testes are the male sex organs which make **sperm**.

Semen is the mixture of sperm and liquid which leaves the **penis**.

9e From girl to woman

Girls often begin puberty slightly earlier than boys. Their body shape changes and their reproductive organs prepare for pregnancy.

Puberty in girls usually begins when they are between 10 and 15 years old. As in boys, it may happen quite quickly or it may take several years. Although it is different for everyone, and everyone ends up a slightly different shape and size, all girls share the same basic changes.

Female reproductive organs

The female reproductive organs have two main jobs. They produce the female sex cells and they provide a home for a growing baby. When a woman is 45 to 55 years old, her ovaries stop producing eggs and she can no longer have babies.

An egg leaves the **ovary** each month. It travels down the **Fallopian tube** to the **uterus** (womb). The uterus is where a baby grows and develops during pregnancy. The passageway leading from the outside of the body to the uterus is called the **vagina**. Sperm come in through the vagina to find the egg, and a baby has to come out through it at birth.

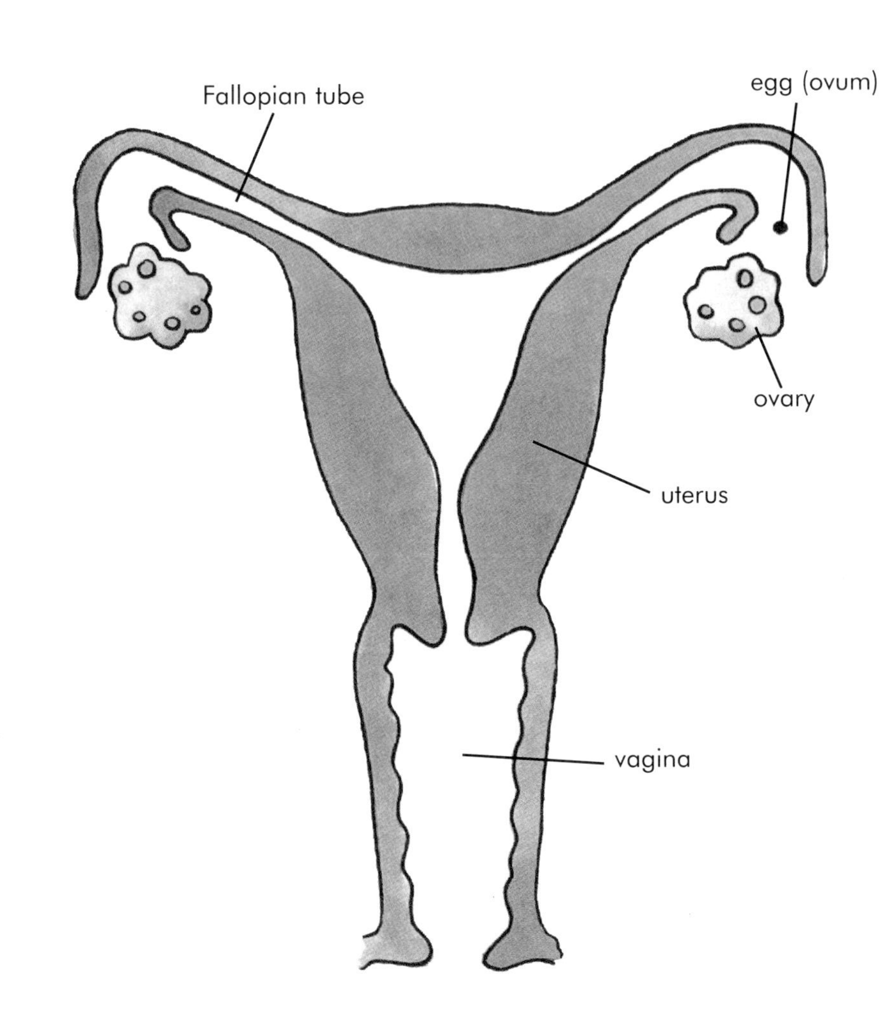

Girls become women

During puberty, a girl's **breasts** develop. They become larger and the nipples become more obvious. The size and shape of the breasts varies a lot from one girl to another, but whether they are large or small most will be capable of producing milk to feed babies.

A girl's body shape alters as fat develops around her hips, bottom and thighs. Along with the growing breasts this gives her a more curvy 'female' shape. Pubic hair appears under her arms and in her groin.

Inside her body her ovaries begin to release a ripe ovum or egg (the female sex cell) each month. Her uterus prepares for pregnancy each month so her monthly **period** begins.

A girl also grows quickly during puberty and nearly reaches her adult height by the end of it.

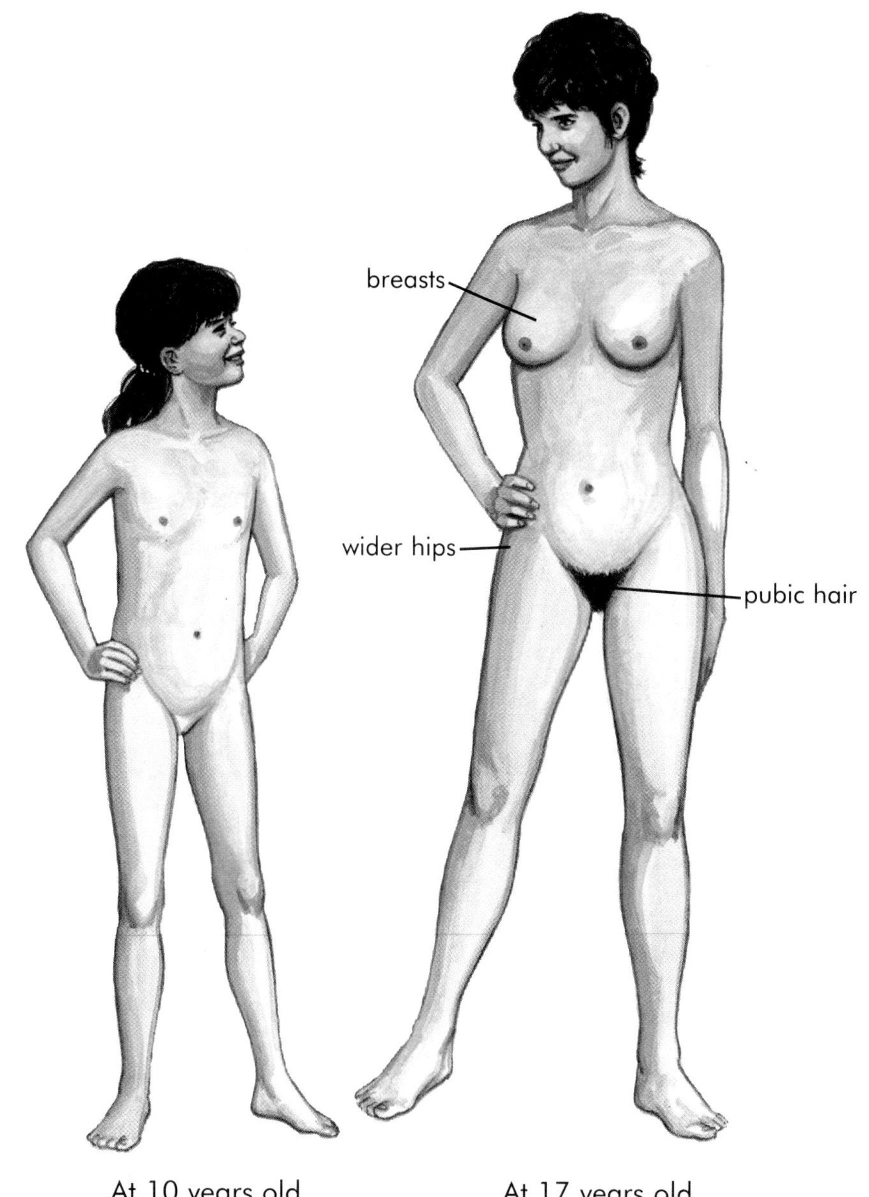

What do you know?

1 Make a list of all the changes that take place in a girl's body at puberty.

Key ideas

During puberty a girl's **ovaries** start producing an ovum (egg) each month.

Her **uterus** prepares for pregnancy each month and she has a monthly **period**.

Her **breasts** develop to prepare for feeding a baby.

9f Meet the stork

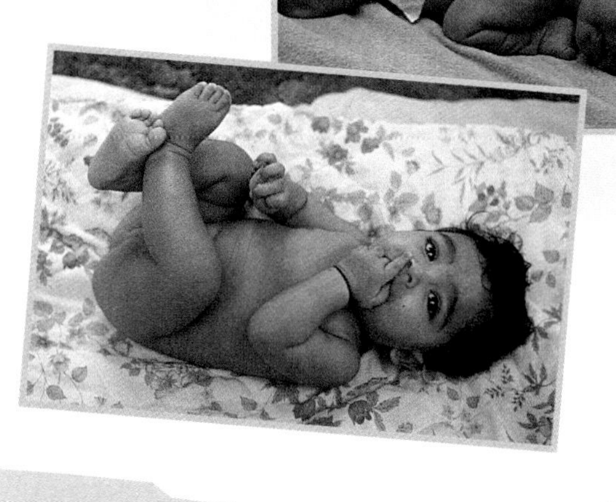

Where do babies come from? Embarrassed parents often tell small children that babies are brought by storks or found under gooseberry bushes, but the truth is even more incredible! How many different explanations have you heard?

Recipe for life

This recipe makes it all sound far too simple, but it does give us the basic facts. In human reproduction, a sperm joins with an ovum (egg) inside the body of a woman. The baby then develops and grows inside her body for about 40 weeks before it is born.

How to Make a Baby
Ingredients:
1 egg
1 sperm
Join together and leave in a warm nourishing place for 40 weeks........

Journey of a sperm

The sperm get inside the woman's body during **sexual intercourse**. This is also known as 'making love' and 'having sex'. It is a very enjoyable part of a loving relationship between a man and a woman.

When the man feels sexually excited his penis fills with blood and becomes stiff (**erect**). When the woman feels sexually excited her vagina becomes wider and very moist. This makes it easy for her partner to slip his penis into it.

During sex the man and woman move their bodies against each other which makes them both feel good.

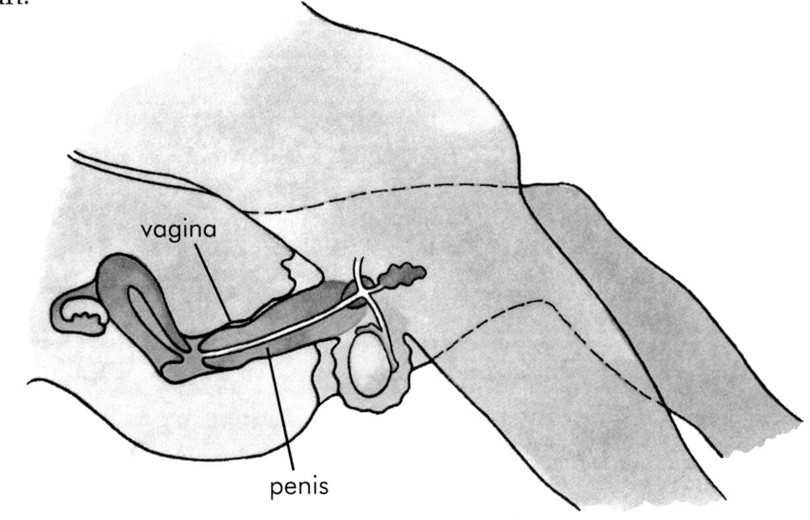

Sperm meets egg

When the man is very excited he **ejaculates**, which means that his semen is pumped out through his penis into the woman's vagina. Although only a small amount (about a teaspoonful) is produced each time, it contains about 500 million sperm.

The sperm can easily swim in the semen up into the uterus and on into the Fallopian tubes. Millions of sperm die on the way, but if an ovum is passing down one of the Fallopian tubes at the time, one sperm may join with it and fertilise it.

When a sperm gets into an ovum, a new cell is formed. This is a new human life.

A baby on the way

If fertilisation happens, the sperm and egg form a single new cell. A baby contains millions of cells, so a lot of developing needs to happen. The single cell will grow and divide, then the new cells will divide again. This will happen for about 40 weeks until the fertilised ovum has become a baby ready to be born.

What do you know?

1 Copy and complete the following sentences. Use the words below to fill the gaps.

ejaculates **penis** **uterus** **sperm** **vagina** **ovum** **sexual intercourse**

To make a baby, an ________ and a ________ must join together. They are brought together during ________ ________. The man's ________ becomes erect and is placed inside the ________ of the woman. When the man ________, millions of sperm swim up through the ________ and on into the Fallopian tubes.

2 The sperm do not always meet an ovum. Why not?

3 Why do you think so many sperm are produced at a time, but usually only one ovum at a time?

Key ideas

During **sexual intercourse**, a man inserts his penis into the woman's vagina.

When he **ejaculates**, sperm are released into the vagina. They swim up into the uterus and Fallopian tubes.

If a sperm meets an ovum (egg), fertilisation takes place.

9g What's happening inside?

The menstrual cycle

Once a girl is well into puberty, her **menstrual cycle** begins and her **periods** start. What is actually happening?

1 Each month some eggs start to ripen in the ovaries. After about two weeks one ovum (egg) will be ripe and ready to leave the ovary. In the uterus, a special rich lining builds up of little blood vessels, cells and mucus. This is ready to receive the egg if it is fertilised and support the developing baby.

2 The ovum pops out from the ovary into the Fallopian tube. This is called **ovulation**. While the ovum is in the Fallopian tube, it could be fertilised by a sperm and begin to grow into a baby. The ovum travels towards the thick lining of the uterus.

3 If the ovum is not fertilised, it dies. The lining of the uterus isn't needed, so about two weeks after ovulation the body gets rid of it. This lining is lost through the vagina as the period. This lasts just a few days, and the whole cycle starts again.

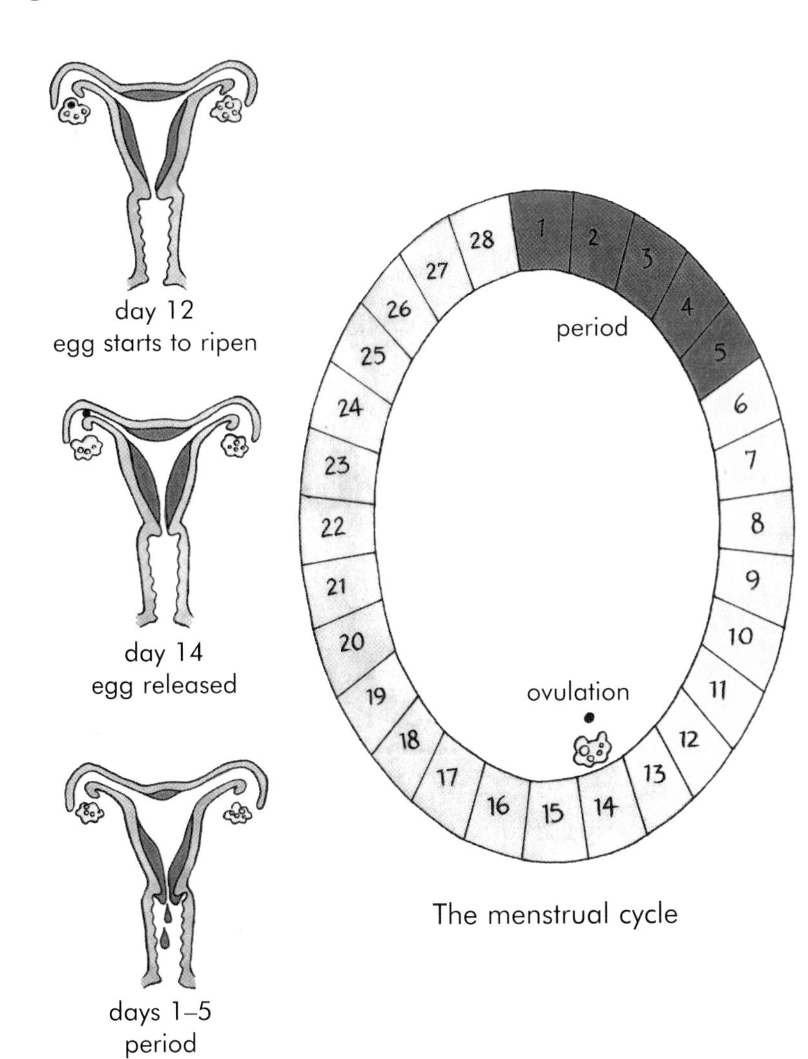

The menstrual cycle

The menstrual cycle starts during puberty. It takes about 28 days (four weeks) from beginning to end, although this varies a lot, particularly when a girl's periods first start. The menstrual cycle stops when a woman is about 45 to 55 years old. This is called the **menopause**.

A nice place to settle

Most months, the egg is not fertilised. Even if the woman has sexual intercourse, the sperm and egg may not meet.

But when an egg is fertilised, it travels on through the Fallopian tube and nestles into the rich lining of the uterus. This is called **implanting**.

Double trouble

Usually people have one baby at a time. Sometimes two babies, known as twins, are born. Where do twins come from?

Sarah and Emma are **identical twins**. One sperm from their father fertilised a single ovum from their mother. Instead of growing into one baby as usual, the fertilised egg split into two as it started to develop, and so two identical babies grew. No one is sure why this happens.

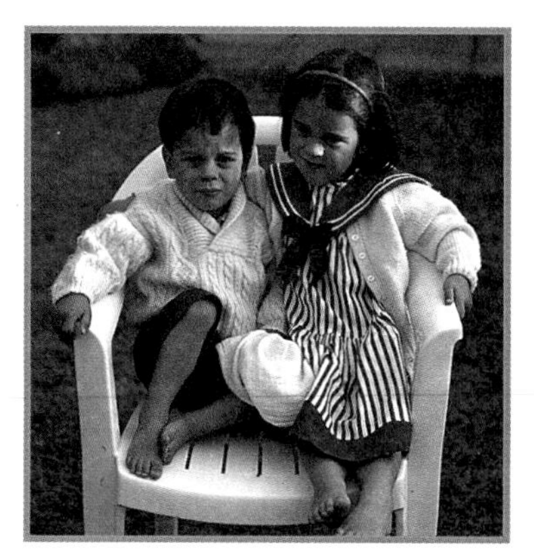

Jack and Julie are **non-identical twins**. Two ova were released by their mother's ovaries, and each ovum was fertilised by a different sperm from their father. Although they were born at the same time, they are no more alike than any other brothers or sisters.

What do you know?

1 Copy and complete the following sentences. Use the words below to fill the gaps.

uterus	**28**	**period**	**Fallopian tube**
develop	**egg**	**fertilised**	**ovary**

The menstrual cycle lasts for about _______ days. An _______ ripens in the _______ and is released into the _______ _______ after about two weeks. At the same time the lining of the _______ builds up. If the ovum is _______ it will implant itself in the lining to _______. If the egg is not fertilised the lining is lost as the monthly _______.

2 Why doesn't a woman have periods when she's pregnant?

3 Explain why:

a some twins look exactly the same as each other

b some twins look completely different and can even be different sexes.

Key Ideas

The **menstrual cycle** lasts about 28 days. An egg ripens and leaves the ovaries each month. The lining of the uterus builds up. If the egg is fertilised it will **implant** into this lining. If not, the lining is lost as the **period**.

Identical twins form when a single fertilised egg splits in two. **Non-identical twins** form from two eggs and two sperm.

9h The end and the beginning

When a woman has a baby growing inside her uterus, we say she is **pregnant**. Inside every pregnant woman, a new person is growing and moving.

a Think about what sort of things a baby in the uterus might need.

All mod cons . . .

As the fertilised ovum grows, it is called the **fetus**. At first it forms a ball of cells. This develops into a tiny human being with its own life-support system called the **placenta**.

The placenta forms at the same time as the fetus. The mother gives food and oxygen to the fetus through it, and takes away waste products. The placenta also protects the fetus from most diseases and harmful substances in the mother's body (but not all).

The **umbilical cord** connects the fetus to the placenta and carries things between the two. Your 'tummy button' is where your umbilical cord connected you to your mother.

A **bag of fluid** cushions the fetus from knocks and bumps. It also makes it easy for the fetus to move about, and they certainly like to wriggle and kick!

By the end of the first 12 weeks of pregnancy the fetus has all its organs. For the rest of the time it is mainly getting bigger and more mature. By about 28 weeks of pregnancy it stands a good chance of surviving if it was born. Towards the end of the pregnancy its head usually points downwards, as this is the best way to be born.

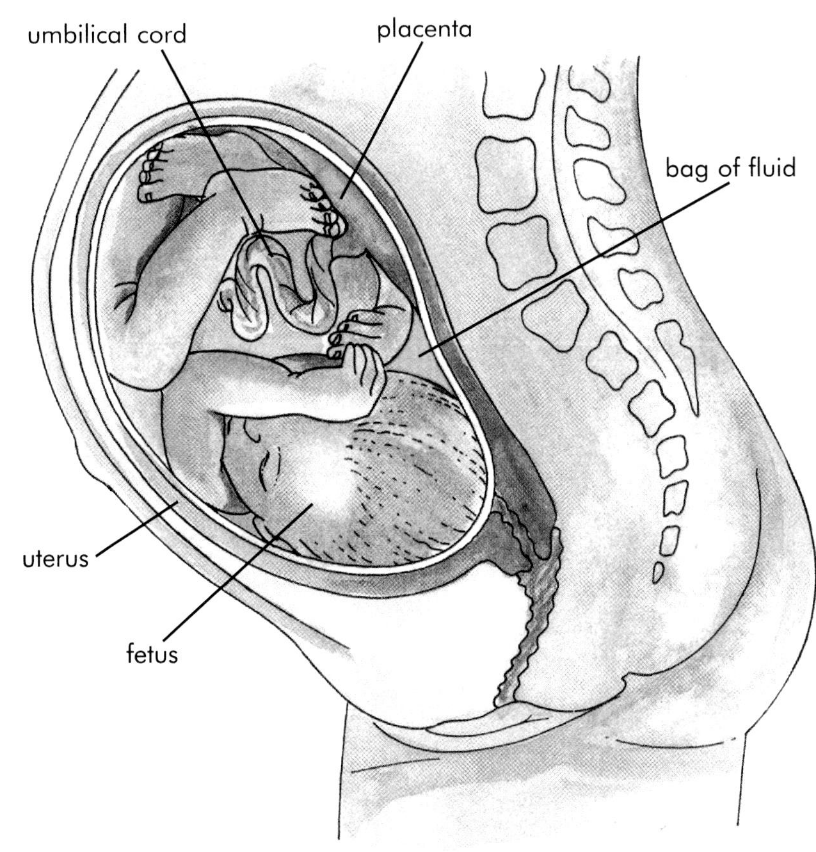

No more room

By the end of a 40-week pregnancy the fully grown fetus is getting too big for its mother's body. It's time to get out!

The baby is pushed out of the uterus during **labour.** This can take several hours. The entrance to the uterus opens up and then the baby is pushed out, normally head first, through the vagina. This is very hard and often painful work for the mother. Once the baby is born, the umbilical cord is cut and a new life has really begun. Finally the placenta also comes away, its job over now the baby is born.

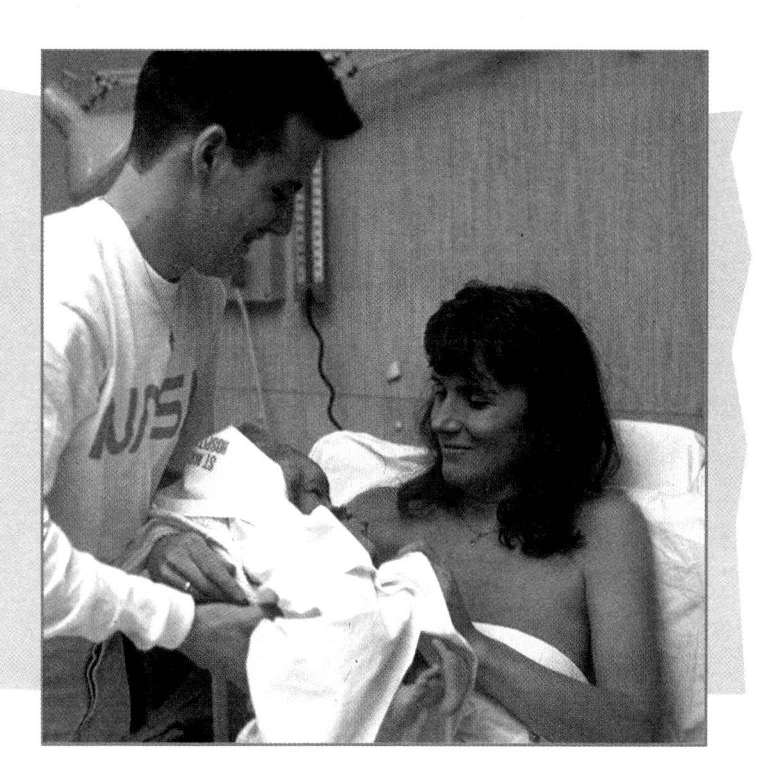

A new beginning

Although birth is the end of pregnancy, it is just the start of a new life for the baby and its parents. Human babies are very helpless. They need milk from their mother's breast or from a bottle, they need to be kept clean and warm and most of all they need to be loved.

What do you know?

1 Copy and complete the following sentences.

a A human pregnancy lasts for about _______ weeks.

b The developing baby is known as a _______.

c The _______ supplies food and oxygen for the developing baby, and also removes all the waste products.

d The fetus is supported in a _______ _______ _______ which protects it from knocks and bumps.

e The fetus is joined to the placenta by the _______ _______.

f The baby is born at the end of several hours of _______.

g After the baby is born the _______ _______ is cut and the _______ comes away.

2 Either:
Write a paragraph describing what you think life is like inside the uterus for a fetus.

or:
Write a short paragraph from a new baby's viewpoint saying what it feels like to be born.

Key ideas

A **fetus** is a developing baby.

When a woman is **pregnant**, she has a growing fetus in her uterus.

Pregnancy lasts for about 40 weeks.

The process of giving birth is called **labour**.

EXTRAS

9b The cycle of life

Living things start off small, grow and eventually die. So while they are fit and strong, all living organisms need to make more of themselves (reproduce). These different stages in the life of an organism can be arranged into a pattern called the **life cycle**.

1 Draw out life cycles for the following organisms. You will have to think carefully for some of them.

a dog	**a person**	**a tree**	**a frog**

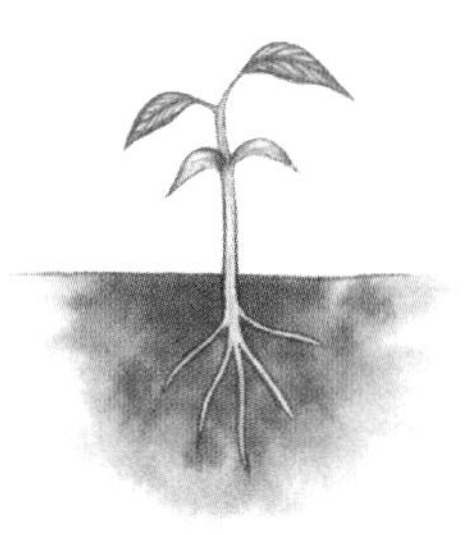

a The young seedling grows from the seed under the soil.

b The plant makes food by photosynthesis and grows big and strong.

c The plant reproduces, making seeds which will grow into new plants. It then ages and dies.

9c More ways of getting around

All over the world there are fruits suited for special ways of dispersal. For example, the tumbleweeds common in the American West use the whole plant to disperse their seeds. Once the fruits have formed, the root of the plant dies and the whole plant blows away in the strong winds, tumbling across the prairies scattering the fruit (and seeds) as it goes.

1 Here are some other strange fruits. Try to work out how they are dispersed.

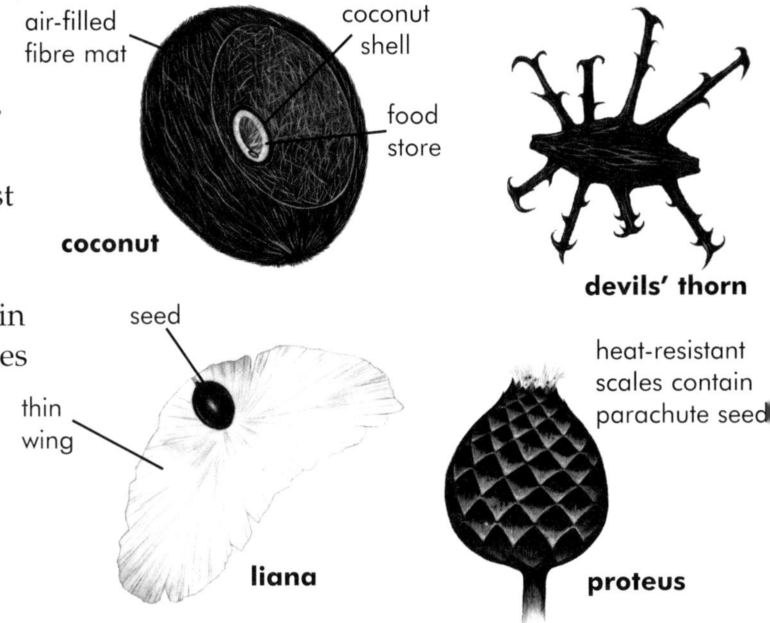

9f The long and short of pregnancy

A human pregnancy lasts about 40 weeks, which can seem like a very long time, but not when you compare it to the pregnancy of an elephant. An elephant is pregnant for about 95 weeks! The length of pregnancy varies greatly between mammals, as this table shows.

Animal	elephant	dolphin	human	house mouse	Virginia oppossum	kangaroo
Length of pregnancy	95 weeks	52 weeks	40 weeks	20 days	12 days	33 days
Number of babies	usually 1	usually 1	usually 1	about 6	12–18	usually 1

1 Why do you think that elephants, dolphins and people have such long pregnancies?

2 What can baby elephants, dolphins and humans do for themselves at birth? How are they different?

3 There are two reasons why animals like the mouse and oppossum can produce babies so rapidly compared with elephants and people. What do you think they are?

4 Kangaroos are very big animals, yet they produce tiny helpless babies after a very short pregnancy. How do the babies survive?

9h It's cold outside

Some babies may be born early (**premature**), sometimes after only 25 or 26 weeks inside their mother's body.

Life is very difficult for these babies. Their lungs are not developed properly so they find it difficult to breathe and they often simply forget to breathe. They haven't got much body fat and their skin is very thin so they lose heat easily and get cold.

Feeding can be a problem for a premature baby. They can digest food, swallow and breathe, but they find it difficult to do all three at once. They either choke and milk gets into the lungs which can be very dangerous, or they don't breathe.

They are also very sensitive to light or sudden noise.

To help them survive, we place them in special cots called incubators. These try to do some of the jobs that would normally be done by the body of the mother.

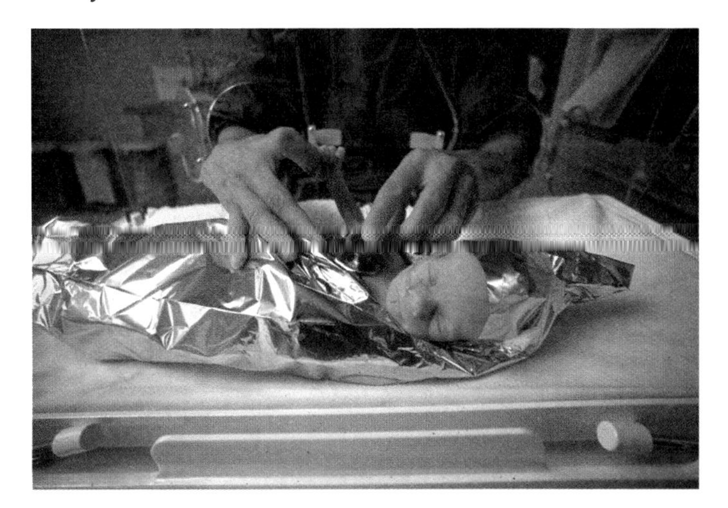

1 Make a list of the things that the incubator, or the nurse caring for the baby, might need to do to help the baby survive.

Switching on

Electricity plays a big part in our lives. It is ready to be used at the flick of a switch.

a Here are three switches. Can you say where each might be found? What might it switch on?

Fire! Fire!

This switch sets off a fire alarm. You have to break the glass with the little hammer so that you can reach the button. The button is a switch connected by long wires to the fire station. An alarm sounds in the fire station to tell the fire crew there is a fire.

How do switches work?

All switches work in the same way. When you press them, they complete an electric **circuit** and make the light or television or radio work.

This picture shows how a switch and circuit work. The switch is made of springy metal. When you press the switch, the bulb lights up. If you take your finger off, the bulb goes out.

Electric current can flow through metal. That is why the switch and the connecting wires are made of metal. When the switch is closed, there is a **complete circuit** of metal for the current to flow around.

Put your finger on the picture of the circuit. Starting at the left hand end of the battery (where the + sign is), trace around the circuit with your finger, all the way to the other end of the battery. This is the path the electric current follows to make the bulb light up.

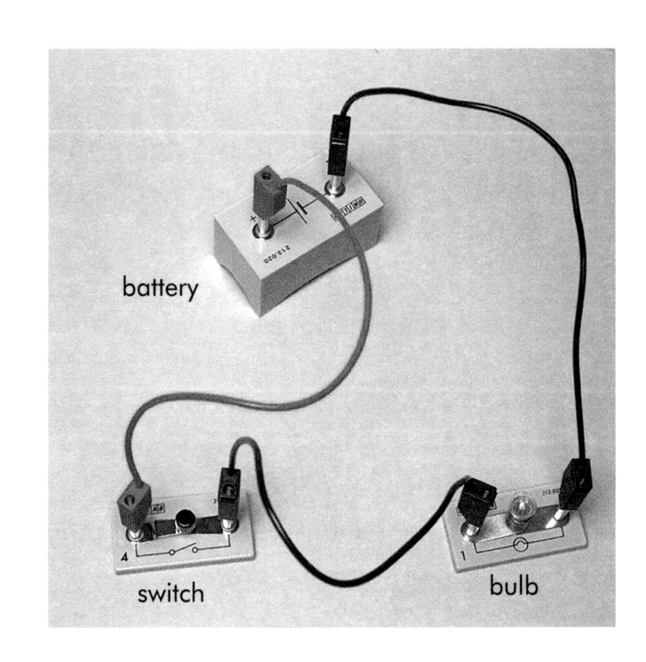

Flowing through

An electric current can flow through metal, and so we say that the metal is an **electrical conductor**. There are some materials that an electric current cannot flow through. These are called **electrical insulators**. Plastic and wood are examples of electrical insulators. (See Unit 3.)

The picture shows an electrical circuit being used to find out which materials are electrical conductors and which are electrical insulators. The pupils test each material by connecting the two clips to it. With some materials, the bulb lights up. This shows that an electric current is flowing round the circuit. With other materials, the bulb does not light up. The current cannot flow round the circuit.

The table shows some of their results. Copy the table and complete it.

Material being tested	Does the bulb light up?	Electrical conductor or insulator?
copper	yes	conductor
cardboard	no	insulator
steel	yes	
polythene	no	
glass	no	
aluminium		
wood		

What do you know?

1 Copy and complete the following sentence. Use the words below to fill the gaps.

circuit **current** **switch**

When a ________ is closed, it makes a complete ________ so that an electric ________ can flow.

2 A connecting wire is made of two materials. It has a metal wire in the middle and a plastic covering on the outside.

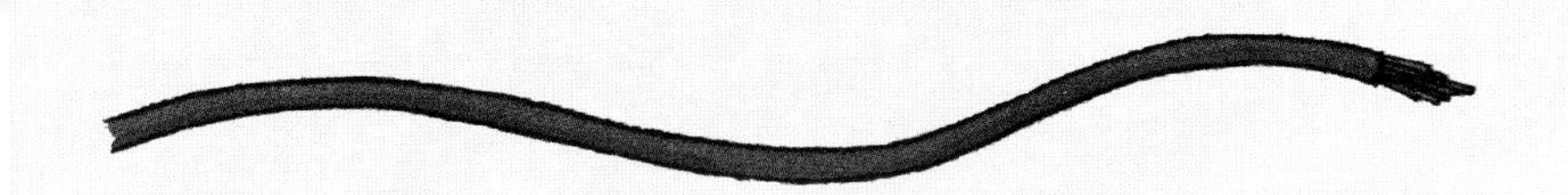

a Which part does the electric current flow through?
b Why is the outside made of plastic?
Use the words **electrical conductor** and **electrical insulator** in your answers.

Key ideas

A switch is used to complete an electric **circuit**.

An **electric current** flows around a complete circuit.

A material that allows an electric current to flow is called an **electrical conductor**.

A material that does not allow an electric current to flow is called an **electrical insulator**.

Wiring up

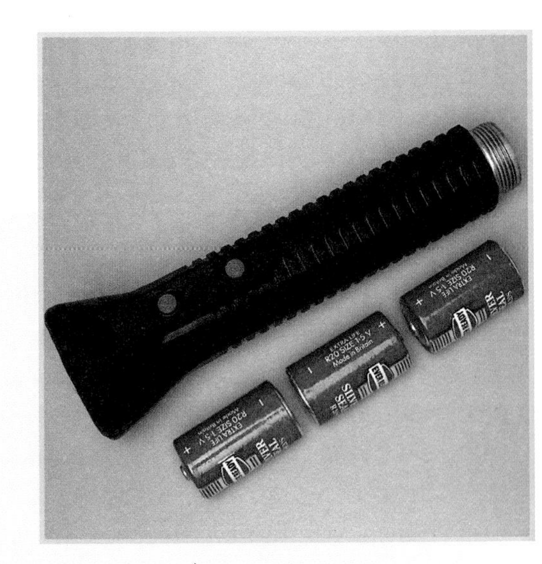

Many electrical appliances need batteries to make them work.

a What appliances do you have that use batteries?

There is always a right way to put the batteries in.
The current flows from positive to negative.

b What is wrong with the arrangement of batteries in this torch?

More switches

A circuit can have more than one switch. Circuits A and B both have two switches.

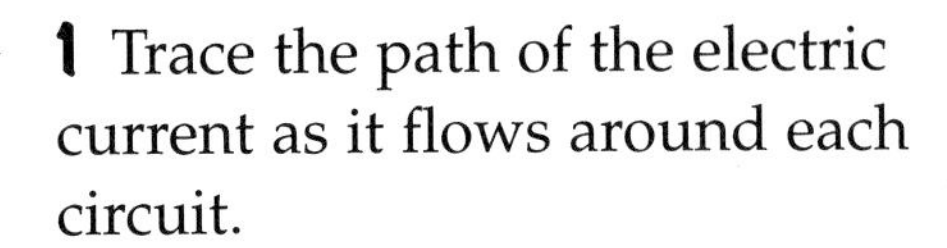

c **1** Trace the path of the electric current as it flows around each circuit.

2 In circuit A, how would you make the buzzer sound?

3 In circuit B, how would you make the buzzer sound? How would you make the motor work?

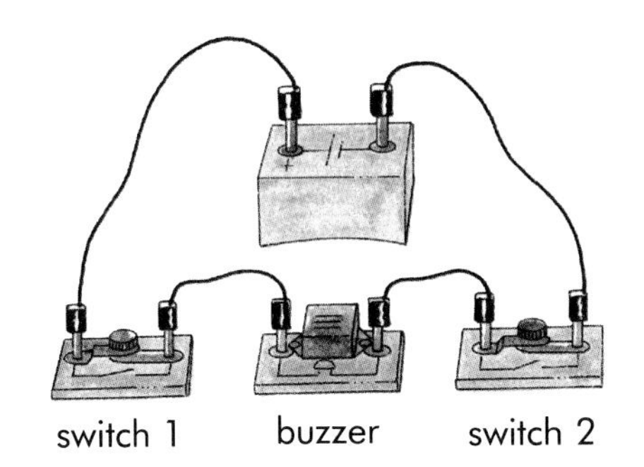

Circuit A

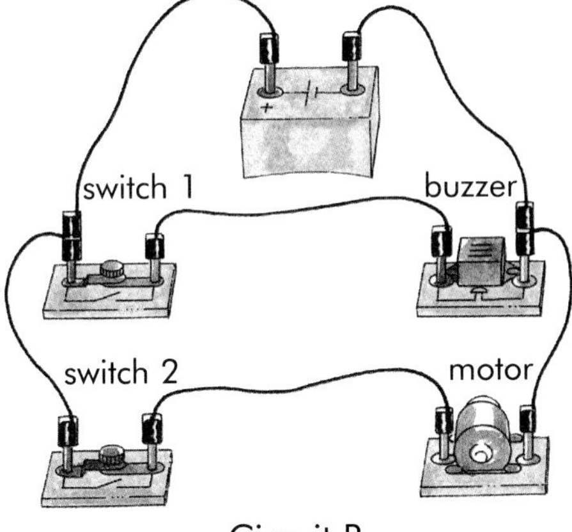

Circuit B

Designing circuits

It isn't very easy to describe an electrical circuit in words. And imagine if the instructions were in a foreign language. It is much easier to use a **circuit diagram**.

There is a **circuit symbol** for every electrical **component** in the circuit. A circuit diagram uses these symbols to show how the different components are connected together.

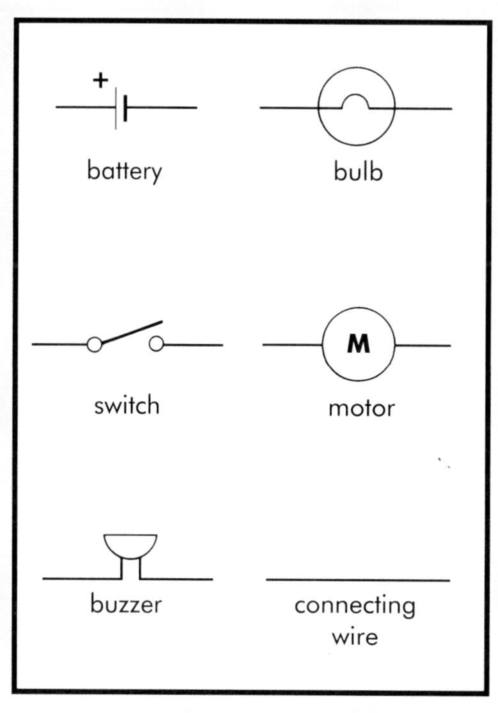

Circuit symbols

This circuit has a battery, a switch and a bulb. You can see how the diagram shows a complete circuit. You can trace the path of the electric current as it flows all the way round the circuit.

This is one of the circuits on the opposite page. Can you say which one?

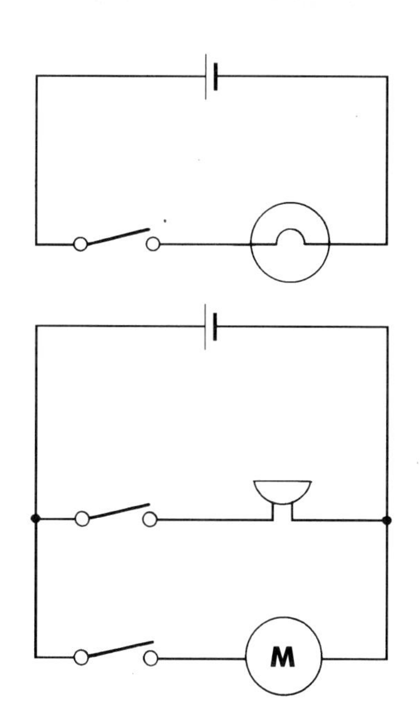

What do you know?

1 The names and symbols for different components have been muddled up in this table. Copy it out, putting the correct name next to each symbol.

Component	Symbol
battery	
bulb	
motor	
switch	
buzzer	
connecting wire	

2 a Draw a diagram to show a circuit with a battery, a switch and a buzzer.

b Now draw a circuit diagram for circuit A on the opposite page.

3 The wires in this diagram are all tangled up. Trace the path of each circuit, and say what will happen when each switch is closed.

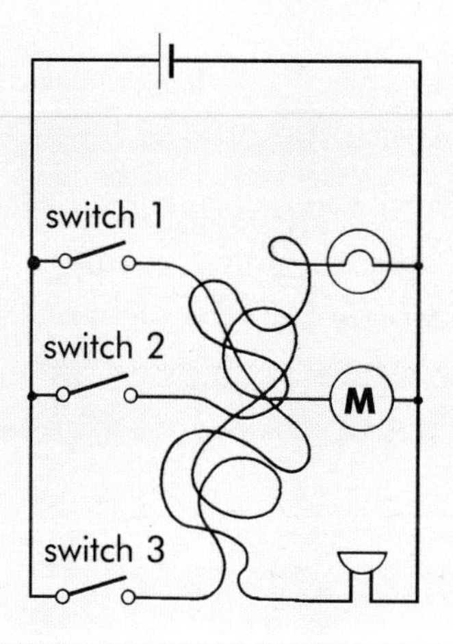

Key ideas

Electric current flows from the **positive** end of a battery, around a circuit, to the **negative** end.

Circuit symbols are used to represent electrical **components**.

A **circuit diagram** shows how they are connected together.

10c Seeing the light

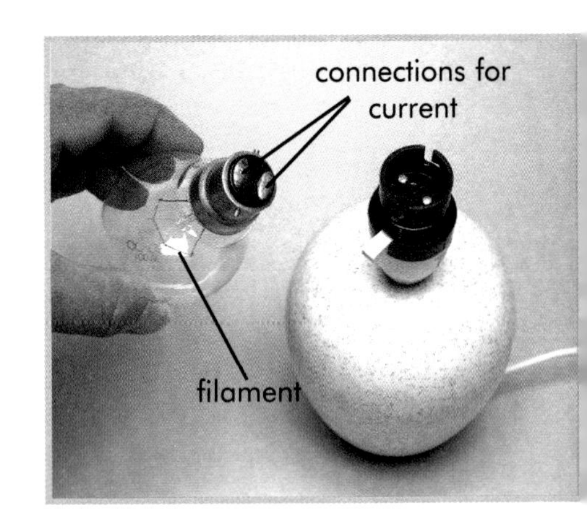

Do you know how an electric light bulb works? Inside the bulb is a curled up metal wire called the **filament**. This gets hot and glows when the current flows through it.

a Why do you think the bulb has two connections in its base?

Brighter lights

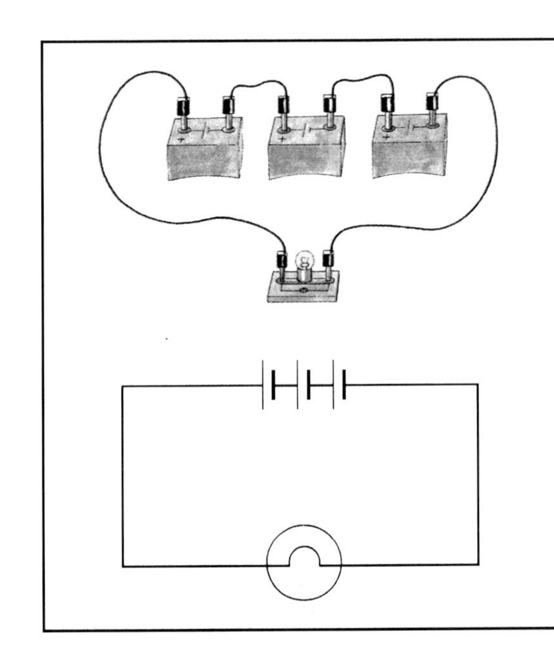

Current flows into the bulb through one connection and out through the other, just as it does through batteries or through the circuit wire.

b **1** You can connect a bulb up to one, two or three batteries. Predict what will happen as you increase the number of batteries.

2 What can you say about the electric current that flows through the bulb when there are more batteries in the circuit?

When several batteries are connected together end to end, we say they are connected **in series**. Batteries in series combine together to give a bigger 'push' to the current in the circuit. A bigger current flows through the bulb, and it lights up more brightly.

Series and parallel

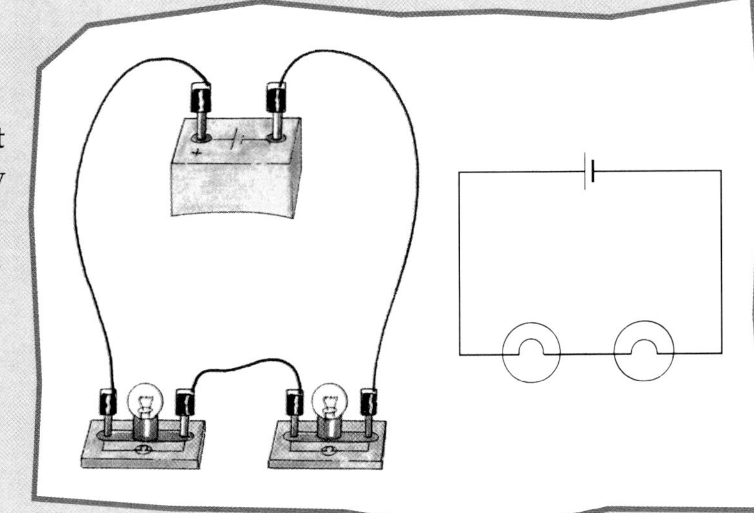

These two bulbs are connected in series. They are one after the other in the circuit. The current flows through one bulb and then the other. They do not light up as brightly as one bulb on its own. The battery finds it more difficult to push the current through two bulbs in series than through one.

c Draw a circuit diagram to show three bulbs connected in series with a battery. Make a prediction about how bright the bulbs will be.

Here is another way to connect up two bulbs to one battery. They are connected side by side. We say that they are **in parallel**. Each one feels the full push of the battery, and they both light up brightly.

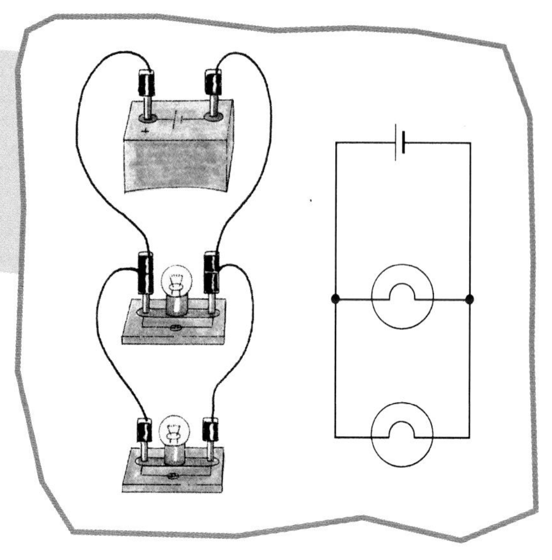

Circuit checker

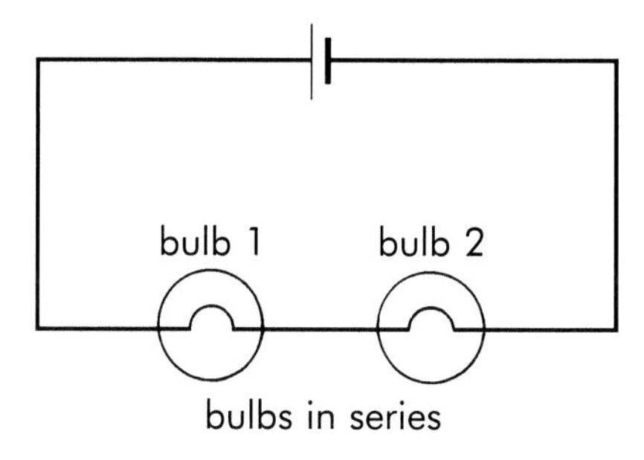

bulbs in series

If you unscrew bulb 1, there is a break in the circuit. The electric current cannot flow round through bulb 2.

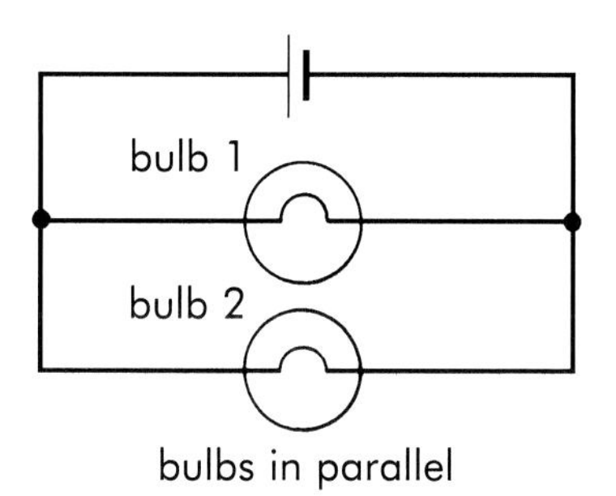

bulbs in parallel

If you unscrew bulb 1, there is still a complete circuit for the electric current to flow round through bulb 2.

What happens if you unscrew bulb 2 instead of bulb 1 in the parallel circuit? Trace the circuit with your finger and write down your ideas.

What do you know?

1 Copy these circuit diagrams. Under each, add a sentence to say what it shows. The first has been done for you.

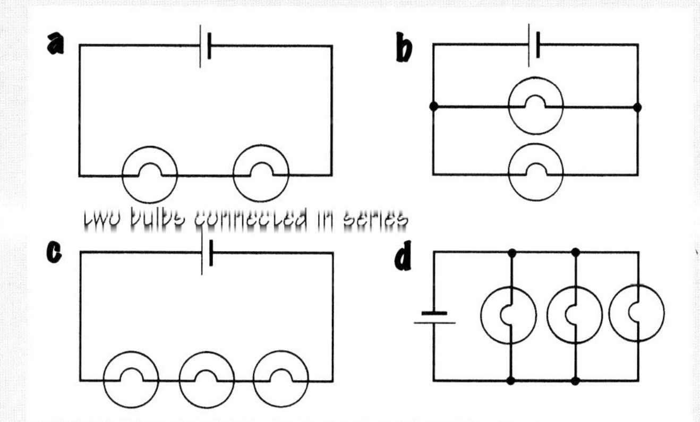

2 a Christmas tree lights can be a problem. When one bulb fails, they all go out. Are they connected in series or in parallel?
b You may have seen a car driving around with only one headlamp working. Are the car's two headlamps connected in series or in parallel?

3 a Explain whether you think the lights in your home are connected in series or in parallel.
b Sometimes there are two lights in a room, and they both work from one switch. How should the lights and the switch be connected together? Draw a circuit diagram.

Key ideas

When batteries are connected end to end (**in series**) they give a bigger push and a bigger current flows.

When two bulbs are connected in series, they shine less brightly than one bulb alone in a circuit.

If you want them to continue to shine brightly, they should be connected **in parallel** (side by side).

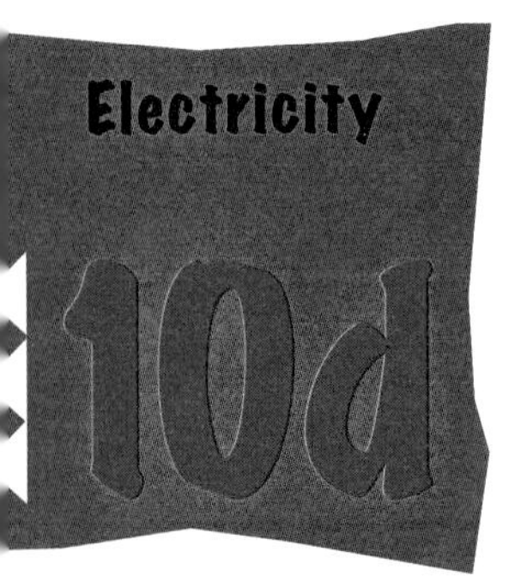

Resistance to change

The lighting technician can change the brightness of the lights by sliding the dimmer controls on a control board. This increases or reduces the amount of current going to the lights.

a What other occasions can you think of when it might be useful to gradually increase or reduce light, or sound, or heat?

Dim it yourself

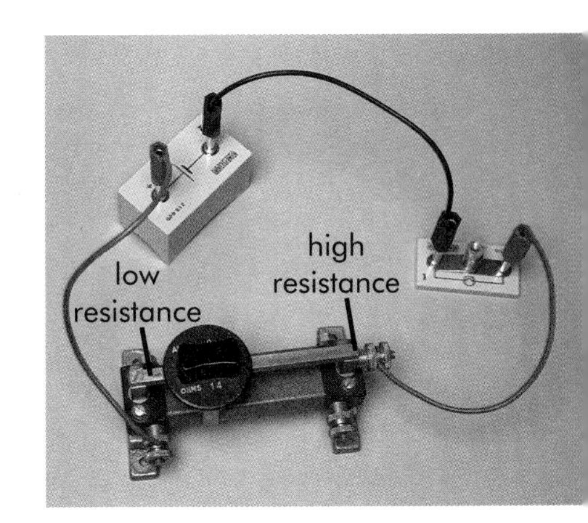

A **variable resistor** works as a dimmer control. Moving the slider changes the current flowing in the circuit. When the current is smaller, the light bulb is dimmer.

A variable resistor works because it has electrical **resistance**. It makes it difficult for the electric current to flow around the circuit. When the slider is at one end, the resistance is low, and it is easy for a big current to flow. When the slider is at the other end, the resistance is higher and only a small current can flow.

The great obstacle race

When an electric current flows around a circuit, it has to flow through all the different components, such as bulbs, motors and variable resistors. It is quite an obstacle race for the current.

In an obstacle race like this, the running children are like the electric current. The obstacles are like the components in the electric circuit.

It is easy to get through a short, fat pipe. It has low resistance. It is much harder to get through a long, thin pipe. It has high resistance.

How can you tell from the picture of the obstacle race that the long, thin pipe has high resistance?

The filament of a light bulb is made of a long, thin piece of wire. It is so long it has to be coiled up to fit it into the bulb. It has high resistance.

DIY dimmer

Pencil lead (graphite) conducts electricity, but it has high resistance. You can make your own dimmer control for a torch bulb using a pencil lead. Slide the crocodile clips along the lead to change the resistance of the lead.

1 Where should the clips be to make the bulb as dim as possible? Where should they be to make it as bright as possible?

2 What happens to the resistance of the pencil lead as the clips are moved further apart?

3 What happens to the current that flows in the circuit as the clips are moved further apart?

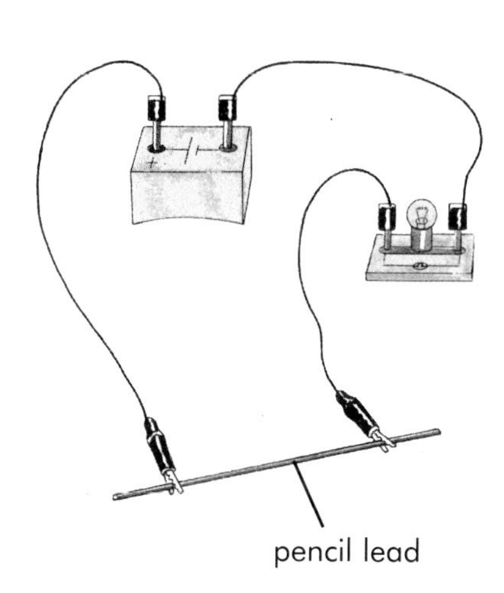

What do you know?

1 Copy and complete the sentences below. Choose the correct word from each pair.

Every component in a circuit has electrical **current/resistance**. The more resistance there is in a circuit, the **greater/smaller** the current that flows. A long wire has **less/more** resistance than a short wire.

2 Which of the following use variable resistors?

a the on-off button on a television
b the brightness control on a computer screen
c the volume control on a radio
d the switch on an electric fire
e the temperature control on an electric oven.

Key ideas

You can change the current that flows in a circuit by changing the **resistance** of the components in the circuit.

With more resistance, the current is smaller.

A long, thin wire has more resistance than a short, fat wire.

How much current?

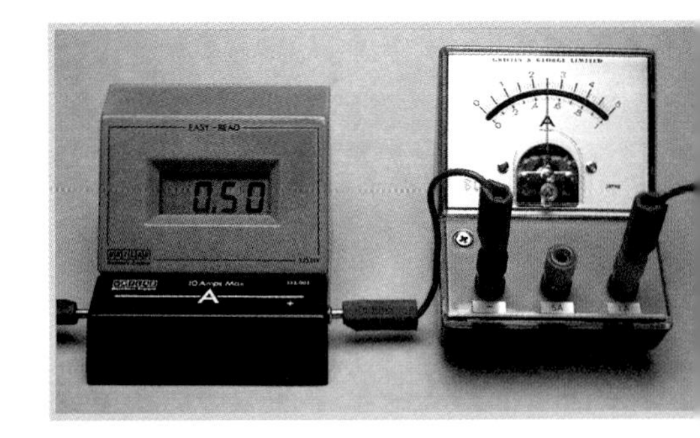

You can measure the current flowing round a circuit using a meter called an **ammeter**. We measure current in **amperes** (A).

The photo shows two different types of ammeter. They both show the same reading, 0.5 A.

This diagram shows how to use an ammeter to measure the electric current flowing round a circuit with a bulb in it. Because we want the current to flow through the ammeter and then through the bulb, they must be connected together in series (end to end).

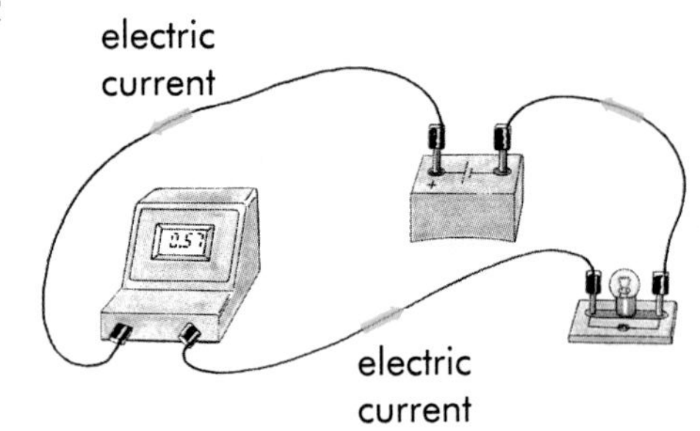

If two bulbs are connected together in series, they shine less brightly than if there was just one bulb.

a What does this tell you about the current flowing in the circuit? How can you check your answer by setting up a circuit?

Measuring the obstacle race

This picture shows the 'current' of children squeezing through the tunnel. The timekeeper is measuring the flow of children to see how many children can get through the tunnel every minute. She is like an ammeter for the obstacle race.

If ten children go into the tunnel every minute, then ten children must come out every minute, even if they have to struggle to get through it. Children don't disappear inside the tunnel! It does not matter whether the timekeeper stands at the beginning or the end of the tunnel. The same number of children will go past her each minute.

The real thing

The tunnel is like the resistance of a light bulb and the moving children are like the flow of current. This suggests that there is as much current flowing out of a light bulb as flows into it. We can check this idea using ammeters.

Both these ammeters show the same reading. There is as much current flowing round the circuit after the light bulb as there is flowing before the light bulb. All the current that flows out of the battery flows round the circuit and back into the battery. Electric current is not 'used up' as it flows round a circuit.

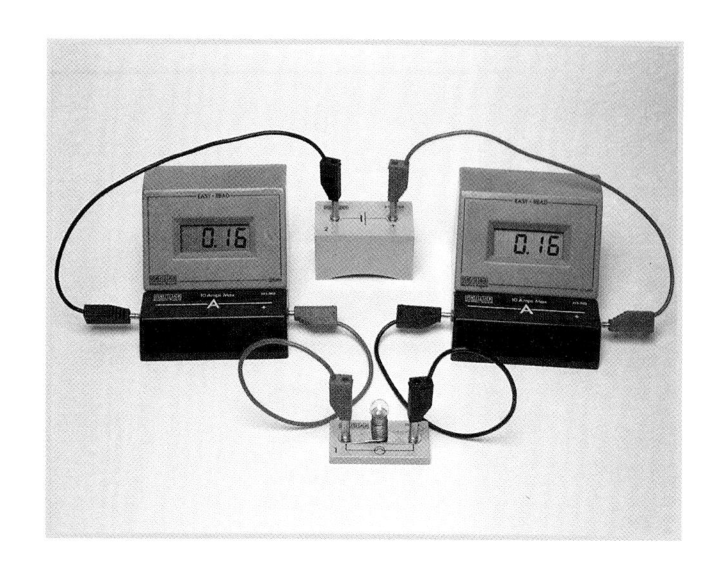

Different appliances have different currents flowing through them when they are working.

1 Which of these needs the biggest current to make it work properly?

2 Which needs the smallest current?

3 How much current flows into the toaster, and how much flows out of it?

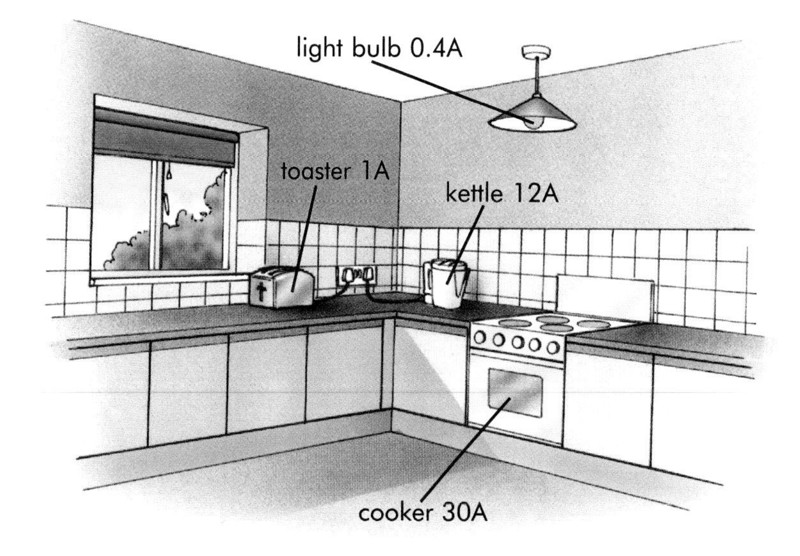

What do you know?

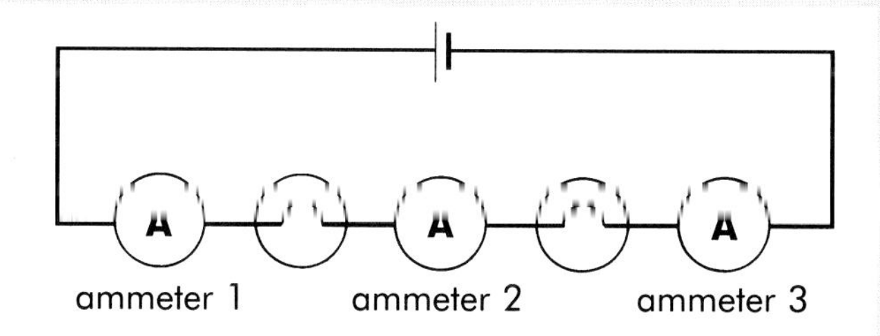

1 In this circuit, ammeter 1 reads 0.5 A. What will the readings on the other ammeters be? Explain your answer.

2 It does not matter whether a switch comes before a bulb or after it in a circuit. Use the idea of electric current flowing around a circuit to explain why this is so.

3 Where would you put ammeters to measure the current flowing in a parallel circuit? Draw a circuit diagram to illustrate your answer.

Key ideas

We use **ammeters** to measure electric current.

We measure current in **amperes** (A).

Current does not get used up around a circuit.

10f Energy changes

Electricity is a useful way of getting energy to the place where you want it. In Unit 4, we looked at ways of using electricity to supply energy. Look back at pages 50–51 to remind yourself how this happens.

a **Electrical energy** is being changed into different types of energy in each of these appliances. Decide what type of energy it is changed into in each one.

A current surprise

It's surprising that all the current that flows out of the positive end of a battery flows round the circuit and back in at the negative end. After all, in this circuit, we can see light coming out of the bulb. Isn't the electric current used up in making light?

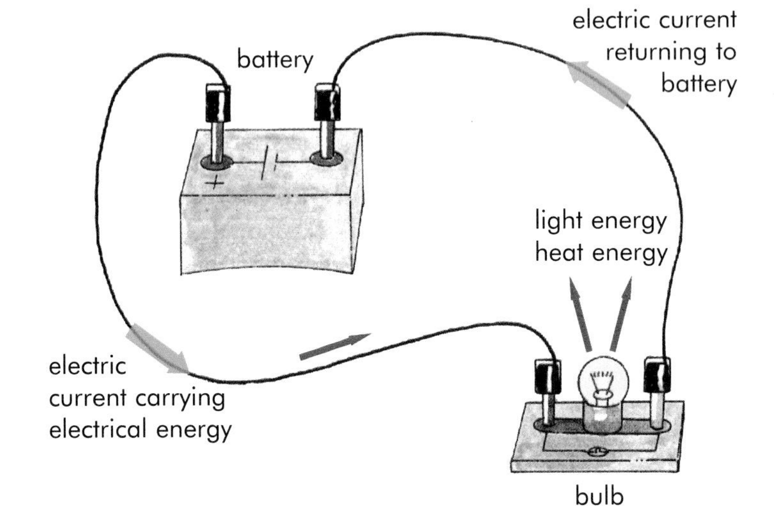

Electric current collects electrical energy from the battery. As it flows, it carries the energy to the bulb. The energy is changed to light and heat in the bulb.

When the current gets back to the battery, it has run out of energy. It collects more energy from the store of chemical energy in the battery before it sets off again.

The same thing happens in the obstacle race. The runners need energy to climb over and through the obstacles. By the time they have finished one lap of the track, they have run out of energy. They need a drink to give them the energy to carry on round again.

Right or wrong?

It is easy to get wrong ideas about electric current. Discuss these ideas about electric current and energy and decide whether they are right or wrong.

Measuring electrical energy

To measure the electrical energy being supplied to a bulb, motor or other device, we use a **joulemeter.** This measures the number of joules of electrical energy that the current is carrying to the motor.

An **electricity meter** is another kind of joulemeter. It measures the electrical energy used in your home. The more you use, the more the electricity company must supply, and the bigger the bill!

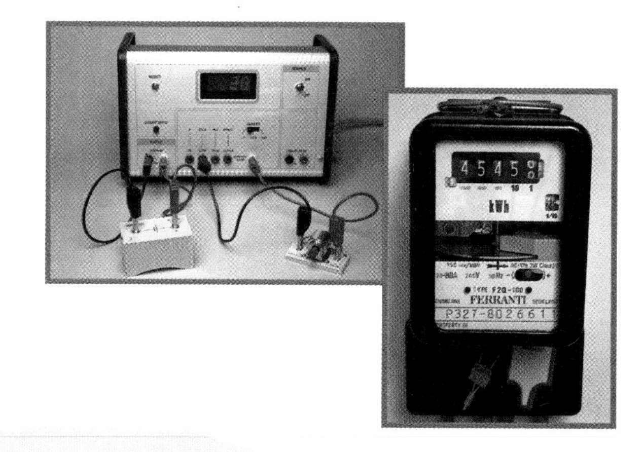

What do you know?

1 Think about the obstacle race model, which is like an electric current flowing round a circuit. Copy and complete the table.

The obstacle race model	A real electric circuit
the obstacles	components in the circuit
	the battery
the moving runners	
the running track	

2 This drawing shows an electrical circuit. A battery is making a buzzer buzz. Copy the drawing and add the labels below at suitable points around the circuit.

a electrical energy carried by an electric current
b chemical energy changed to electrical energy
c current returning to battery with no electrical energy left
d electrical energy changed to sound energy.

Key ideas

An electric current flows all the way around a circuit.

It gains **electrical energy** from a battery, and transfers it to devices which change it into different forms.

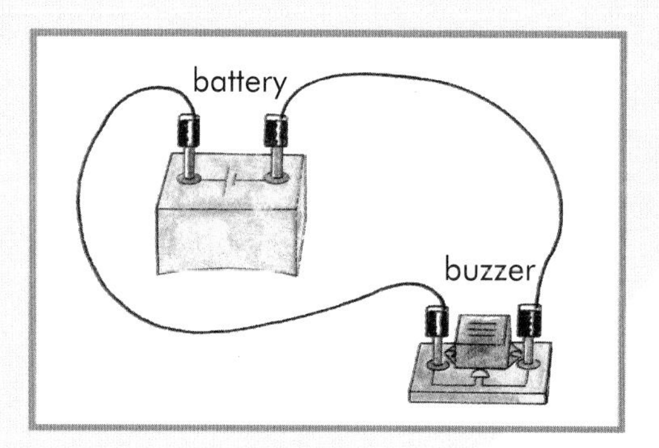

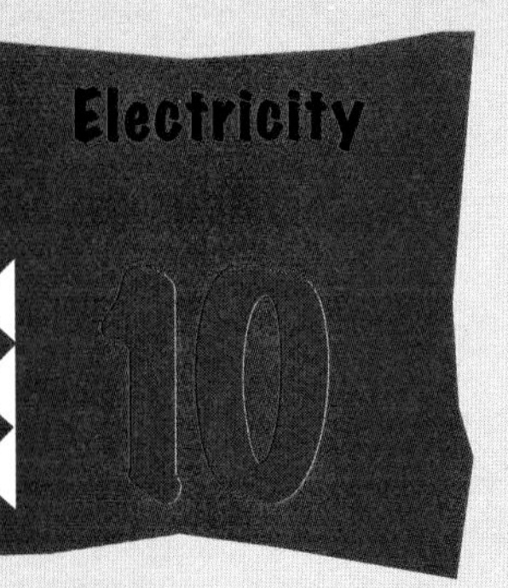

EXTRAS

10a Conductors or not?

Some pupils were doing the experiment to find out which materials are electrical conductors. They used the different materials shown in the picture. They noticed that sometimes the bulb was bright, and sometimes it was dimmer. They made a note of this in their results.

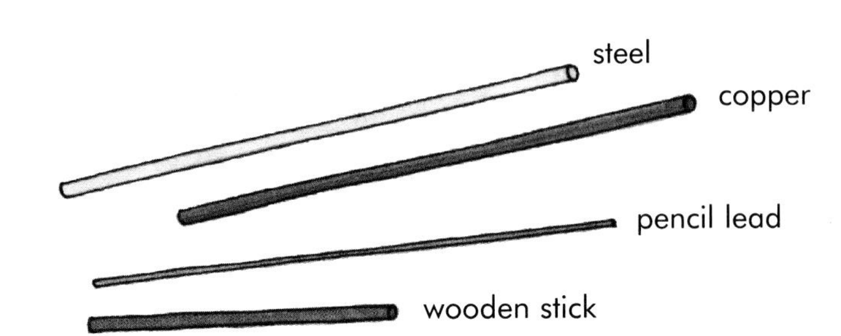

Material being tested	Brightness of bulb
copper rod	very bright
steel rod	bright
wooden stick	no light
pencil lead	very dim

1 What can you say about the different materials the pupils were testing?

2 The pupils should have made their experiment a fair test. How could they have done this?

10b Two-way switches

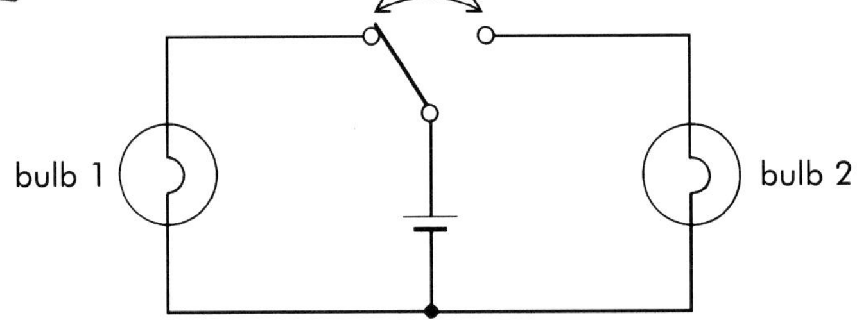

1 This circuit has a different kind of switch. At the moment, bulb 1 is on. What will happen when the switch is changed over?

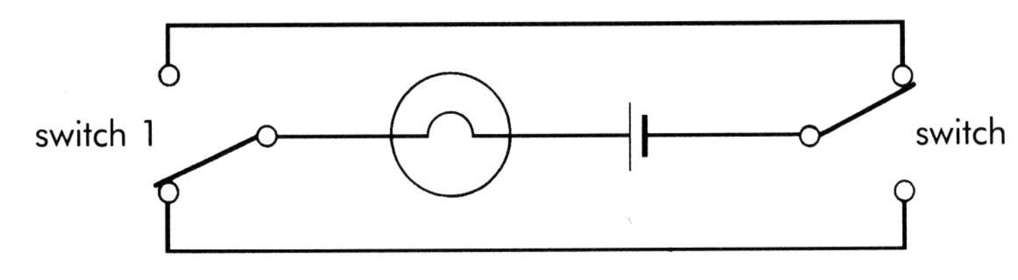

2 This circuit has two switches. With the switches like this, will the bulb be lit up? What will happen if switch 1 is changed? What will happen if switch 2 is then changed?

3 Where might you have a circuit like this at home?

10c Bright lights

1 Which will be brighter, bulb 1 or bulb 2? Why?

2 Which will be brighter, bulb 2 or bulb 3? Why?

3 Which bulb will not be lit at all? Why?

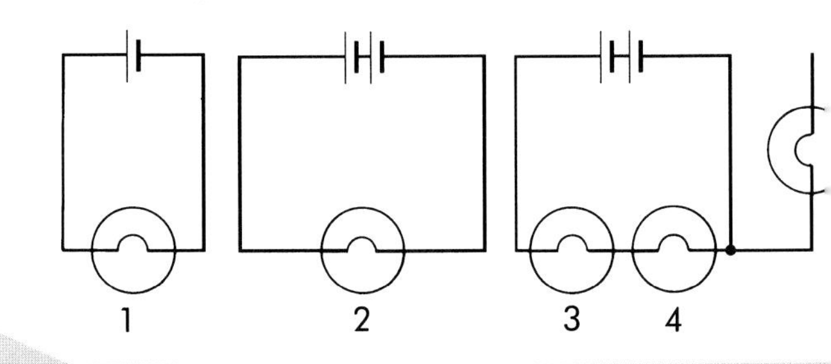

10d Slider circuits

In this circuit, when you move the slider, one bulb gets brighter as the other gets dimmer.

1 When the slider is at the left-hand end, which bulb will be brightly lit?

2 Explain why this bulb gets dimmer as you move the slider, and why the other gets brighter.

3 Where might a circuit like this be useful?

4 Draw a circuit diagram for this circuit.

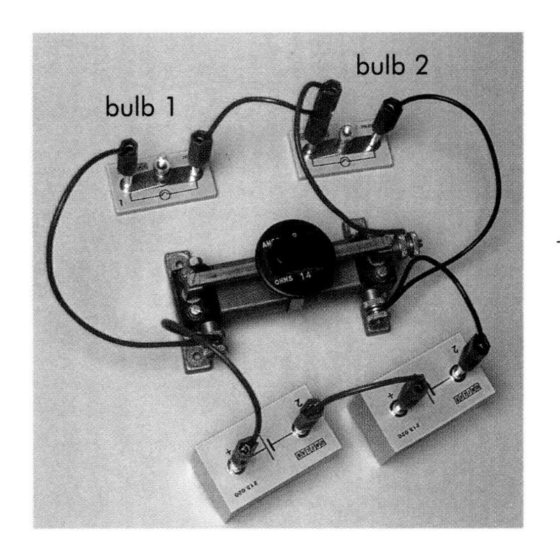

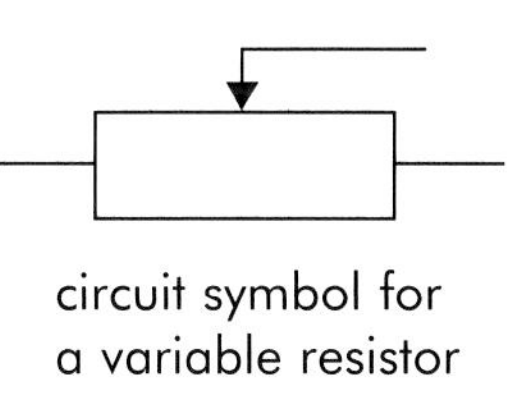

circuit symbol for a variable resistor

10e More obstacles

The obstacle race helps us to understand how an electric circuit works. What do the following ideas about the obstacle race suggest about electric circuits?

1 It is easier to get through a short, fat pipe than through a long, narrow one.

2 If two pipes the same size are placed side by side, twice as many racers can get through in the same time.

3 If two pipes the same size are laid end to end, it will take the racers longer to get round the course.

10f Central heating

An obstacle race is like an electric circuit. So is a central heating system.

1 Write a paragraph to explain how they are similar. Use the words and phrases in the following list.

heat energy	**colder water**
electric current	**complete circuit**
hot water	**flowing**

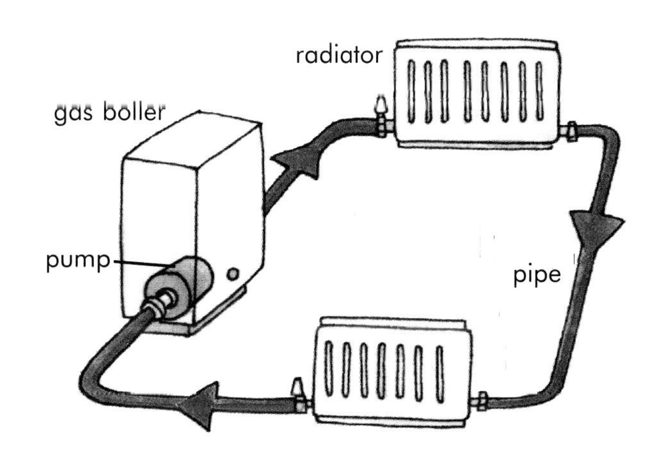

11a In the beginning . . .

When the Earth formed, it got so hot that it melted. As its surface cooled the first rocks formed, giving the Earth a solid crust. Rocks that form from molten rock cooling like this are called **igneous rocks**.

The inside of the Earth is still hot, and its core is still molten.

a How do you think scientists know this?

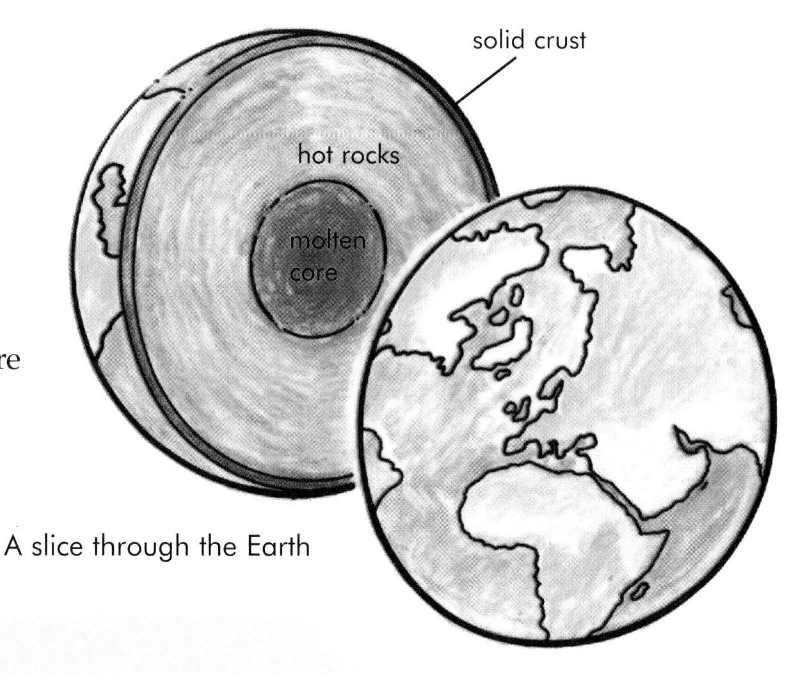

A slice through the Earth

Molten rock

Deep down beneath the crust, the temperature can be 1000 °C or more. Here the rock is a liquid. In places, this forces its way to the surface by cracking the crust, and pours out as **lava**.

This lava cools and starts to set quickly, as temperatures at the surface of the Earth are well below the melting point of rock. The result is that a cone-shaped mound of newly formed rock builds up around the crack. A **volcano** forms.

A volcano has a **crater** at its centre, which is still filled with molten lava. Every few years, the volcano erupts, and more lava flows out of the crater. This flows downhill like a sluggish river, destroying everything in its path.

A volcano called Mount Etna erupting

Looking at igneous rocks

Lava rock is usually a dark brown or black rock. If you look at a thin slice of lava through a microscope, you can see that it is a mixture of tiny crystals. These grew as the lava set.

Gabbro has crystals just like lava rock. But the crystals are much larger. You can easily see them in the rock. These crystals show that the rock is igneous, and formed as liquid rock set.

So why are the crystals so large? Lava cools rapidly on the Earth's surface, so the crystals are small. The liquid gabbro never reached the surface and so cooled much more slowly inside the Earth's crust. This allowed the crystals to grow much larger.

Lava cools rapidly at the surface, so its crystals are tiny. You need a microscope to see them.

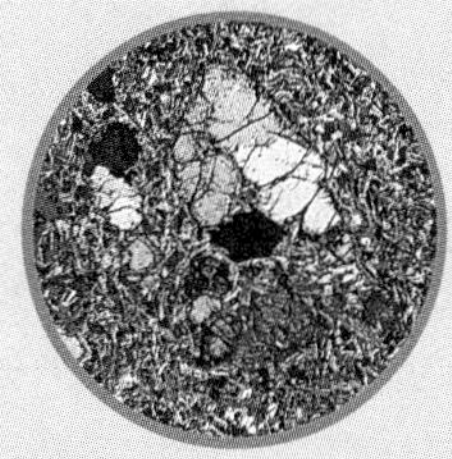

Lava seen under a microscope

Gabbro cooled slowly, deep underground, so its crystals are much larger.

Fast or slow?

These crystals were grown from the same mixture. One lot grew in a few hours, the other took several days.

Which is which? Explain your answer.

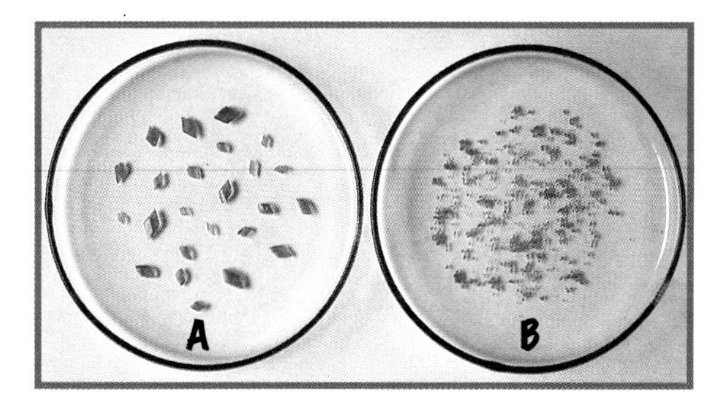

What do you know?

1 Copy and complete the following sentences. Use the words below to fill the gaps.

igneous larger tiny slowly lava volcano

If liquid rock reaches the surface of the Earth, it will often build up a ________. The liquid rock that comes out of a volcano is called ________. When this sets it forms an ________ rock made up of ________ crystals. Gabbro is an igneous rock which has much ________ crystals as it cooled more ________ within the Earth.

2 What evidence is there that it is still very hot below the surface of the Earth?

3 Many people live on and around Mount Etna in Sicily. Why do you think most of them are safe when the volcano erupts?

Key ideas

There is hot liquid rock beneath the surface of the Earth. In places this forces its way to the surface and erupts as a **volcano**.

Rocks that form when liquid rock cools and sets are called **igneous rocks**.

Igneous rocks have crystals which form as the liquid rock cools.

Lava rock has tiny crystals because it cools very quickly.

11b Breaking the rocks

The first rocks were igneous rocks like lava or gabbro. But they no longer cover the Earth. They seem very hard and long-lasting, so where have they gone? The answer is that over millions and millions of years even the hardest rocks get broken up or worn away.

a The picture shows broken-up rocks at the bottom of a cliff made out of lava. How does this happen?

Physical breakdown

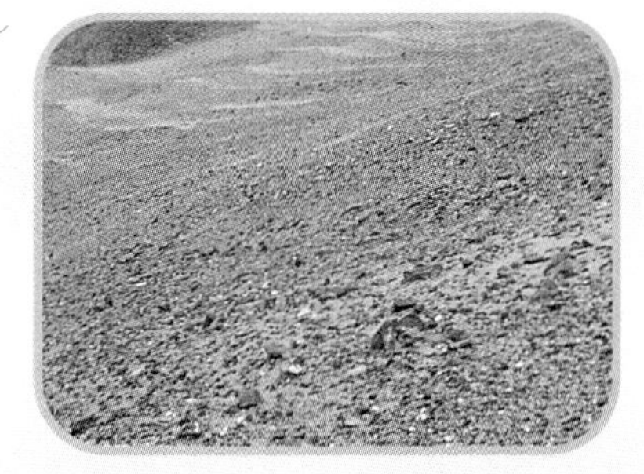

What have these broken rocks got in common with a broken glass in the washing-up bowl?

Most materials expand when they are heated and contract when they cool. This can make things break if they expand or contract too fast or too often. For example, glasses sometimes break if you pour boiling water into them.

In the desert it is very hot by day but can be very cold at night. So desert rocks expand and contract day after day after day. This gradually breaks them up into smaller pieces.

Ice power

One material that does not expand when it's hot and contract when it's cold is water. Water actually expands as it freezes to form ice. This process also breaks up rocks in the mountains.

By day, water runs into cracks in the rock.

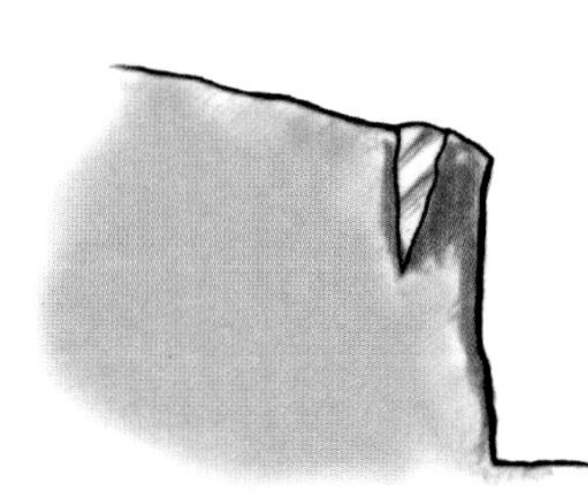

At night, the water freezes, expands and pushes the cracks apart.

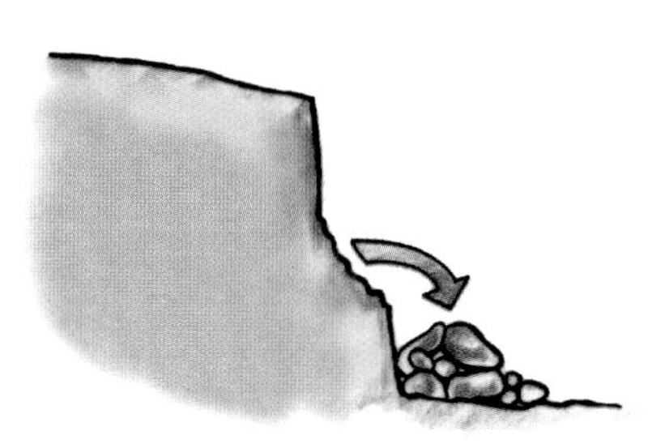

The process can be repeated many times and eventually the loose rocks fall away.

Grinding down

Small pieces of rock get carried along by rivers. They bump and grind along the river bed, crashing into each other as they go. This chips off their corners, so they get rounder and smaller as they roll downstream.

Chemical attack

It is not just **physical** processes like heat and ice that break rock. Rocks are also under **chemical** attack from two highly dangerous chemicals, water and air!

That might seem a little odd, as you need water and air to live. But remember what water and air do to iron. Water and air can gradually turn pieces of igneous rock into sand and clay.

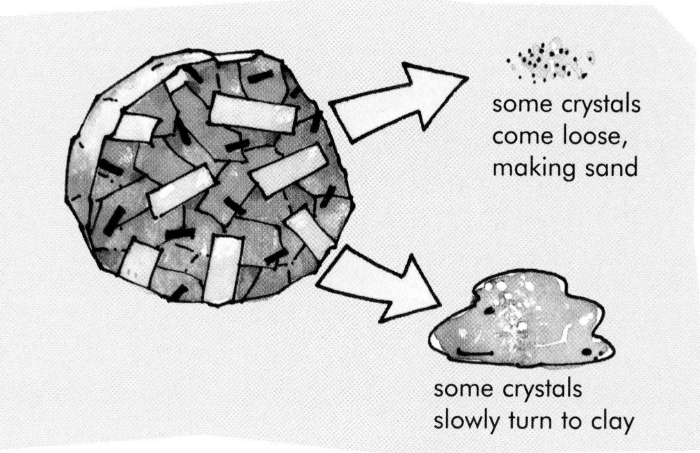

It's all in the soil

When rocks break up and decay like this, it is called **weathering**. The final result is **soil**. Soil is important for plants. Their roots can easily dig into it and grip firmly for support.

- Look back at Unit 5 and find two other important things that plants get from soil.

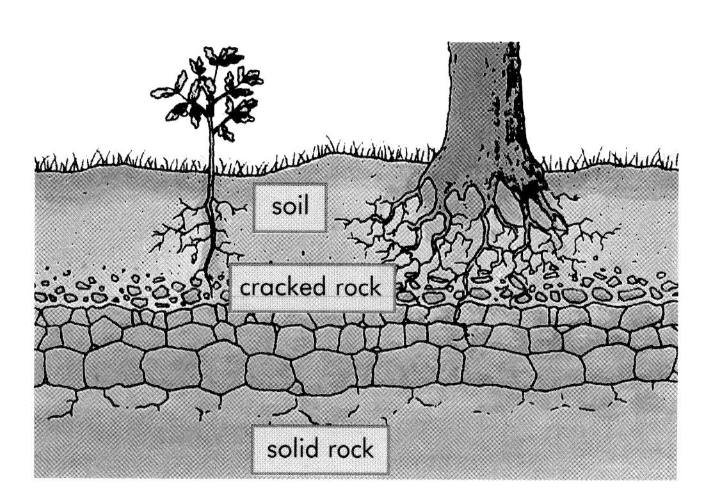

What do you know?

1 Copy and complete the following sentences. Use the words below to fill the gaps.

sand weathering heating soil ice

Igneous rocks are slowly broken into pieces by ________ and cooling. When water turns to ________ it can push cracks open. Air and water then help to break the pieces down into ________ and clay. This process is called ________, and the rocks slowly turn into ________.

2 Deserts are very hot by day and very cold by night. What effect does this have on rocks in the desert?

3 Pipes burst and rocks split when water freezes. What unusual property of water makes this happen?

Key ideas

Rocks break down by both **physical and chemical processes**.

Rocks can be broken by heating and cooling.

Water freezing in cracks expands and pushes rock apart.

Rocks fragments are worn down as they are rolled by rivers.

Soil forms when rocks are weathered.

11c Move it!

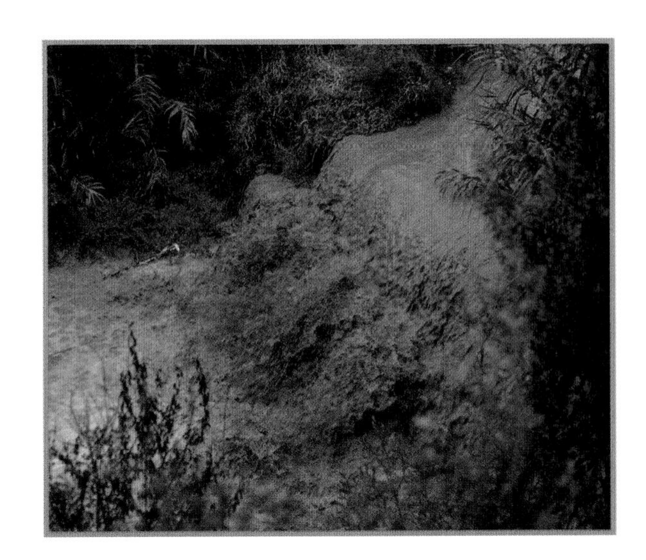

Soil is made from the sand, clay and rock fragments that form when rock is weathered. But heavy rain can wash this material into rivers, and rivers can carry it along as they flow.

Rivers carrying **sediment** like this wear away more rocks as they flow over them. In this way, rivers slowly cut through the land. This process is called **erosion**.

a The water of this fast-flowing river looks brown because of all the clay, sand and rock it is carrying. Where do rivers take all this sediment, and what happens to it?

What can a river carry?

Big rivers can carry more sediment than small rivers. But the speed of the water is also important. A river may carry more sediment in a short period of flood after heavy rain than it does in the rest of the year!

Fast-flowing mountain streams are very powerful. They can carry clay and sand mixed up with the water, and roll pebbles and even boulders along the river bed.

As rivers leave the mountains, the slopes become gentler and the water flows more slowly. Clay particles are very small and stay mixed up in the water. Larger sand grains settle out, but may be rolled along the river bed. Pebbles only move in times of flood.

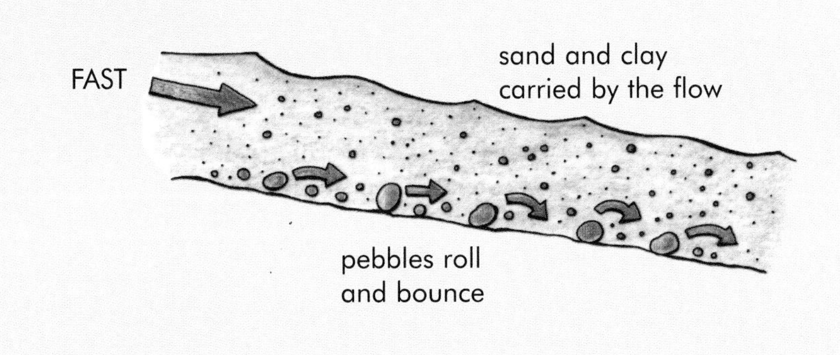

Dump it in the sea

When a river reaches the sea, its power is lost as the waters mix. This lets all the sediment settle to the bottom. If the river is large and is carrying a lot of sediment, it gradually builds a wedge of new land out into the sea. This is called a **delta**.

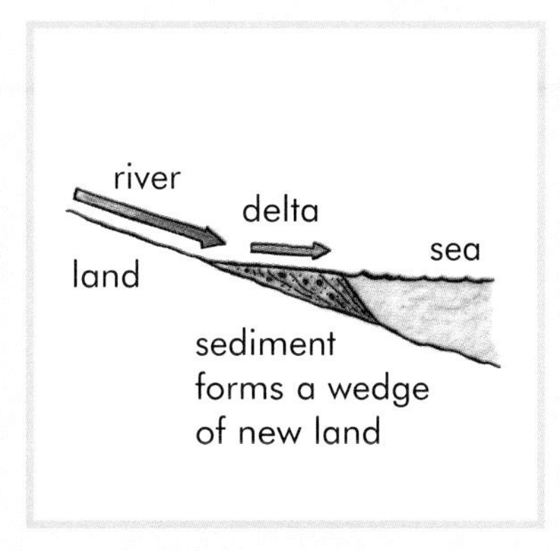

How a delta forms

The Nile Delta seen from space

How do sediments settle?

Jatinder put some pebbles, sand and clay into a screw-topped jar. Ordinary soil usually contains these three types of particles, amongst other things. She filled the jar with water and shook it well. She then let it settle. This is what she saw.

1 The different materials settled out as layers. How many different layers can you see?

2 In what order did the pebbles, sand and clay settle out? Which layer is at the top, which at the bottom?

3 Why did this happen?

4 What does this tell you about the way river sediment settles in the sea?

What do you know?

1 Copy and complete the following sentences. Use the words below to fill the gaps.

sediment flow sea settles

Rivers carry material along as they ________. When they reach the ________, the water slows down so this material ________ out. New layers of ________ are formed on the sea bed.

2 River A seems clear but has a little clay and sand in suspension, and pebbles are rolled along its bed. River B is dark brown from all the clay it is carrying. On its bed, sand and pebbles are not moving. Which river is flowing faster, A or B? Explain your answer.

3 Harbours are enclosed areas of calm water. Why do they often have to have mud and clay removed from the sea bed?

Key ideas

Rivers carry **sediment** along as they flow. The faster they are flowing, the more sediment they carry.

This sediment helps the river to wear away more rocks.

Sediment is dropped as a river enters the sea. This can build up to form new land called a **delta**.

11d New rocks from old

New sediments collect in layers at the bottom of the sea.

In time these beds of soft sediment harden to form **sedimentary rock**. Some sedimentary rocks are so hard that they are used to build houses, or even pyramids!

How does this happen?

Squash it

If you've ever made a clay model, you will know that wet clay is soft and can be easily shaped. Once it has dried out it becomes harder.

Fresh beds of clay contain lots of water, so they are very soft. They cannot dry out and harden under the sea! But they become buried under layers of new sediment. The weight of these new layers above them squeezes out the water. Rocks made from clay like this are called **shale**. Shale is quite a soft rock and easily splits into layers.

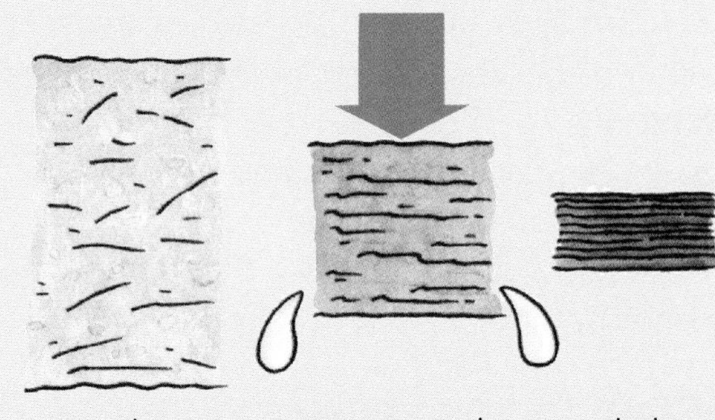

Cement it

If you pour a little water onto a beaker full of dry pebbles, it seems to disappear as it fills up the gaps between the pebbles. The same thing happens with a beaker full of sand, but the sand grains are smaller so you cannot see what is happening so easily.

Sand beneath the sea is always filled with water like this. But the water also contains traces of substances such as lime (calcium carbonate) or silica in solution.

Sometimes these start to form crystals in the spaces between the sand grains. The crystals stick the pieces together, turning the soft sediment into hard sedimentary rock. In this way, sand becomes hard **sandstone**.

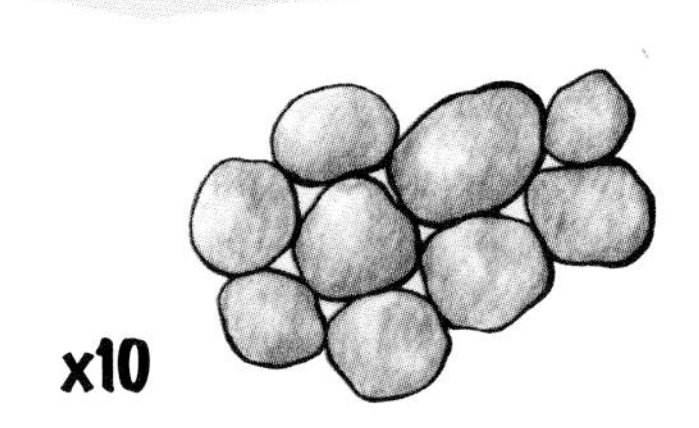

What's the difference?

Andy was looking at some pieces of dry sandstone. Some were very soft and crumbly, while others were very hard. When he dropped some pieces in water, tiny bubbles came out of the sandstone for a while, and then stopped.

- What was in these bubbles and where did it come from?
- Why did the soft sandstone give more bubbles than the hard sandstone?

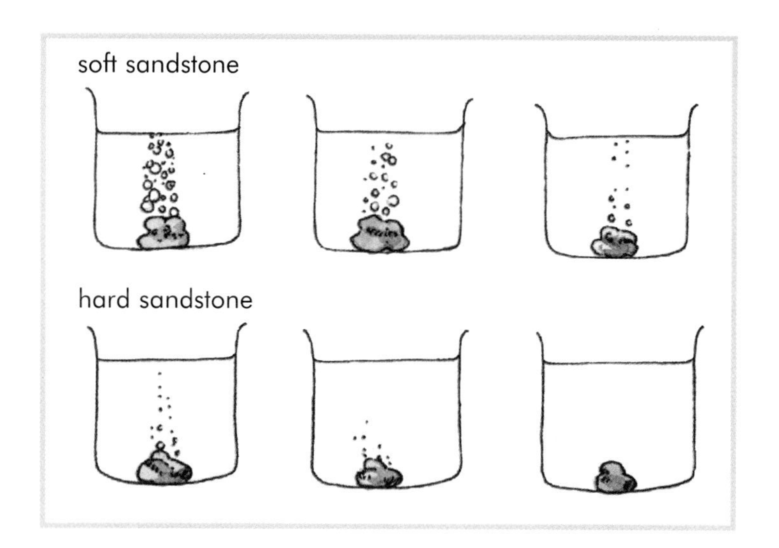

Limestone and fossils

Many animals living in the sea have shells that they grow by taking lime from the water. When the animals die and decay, the shells are left behind. Sometimes rocks are built from the remains of shells. This type of rock is made almost entirely of lime and is called **limestone**. When the remains of ancient living creatures are found in the rocks like this, they are called **fossils**.

A fossil ammonite

What do you know?

1 Copy and complete the following sentences. Use the words below to fill the gaps.

sandstone shale crystals sedimentary

Sediments slowly harden to form ________ rock. Clay turns to ________ when the water is squeezed out. Sand turns to ________ when the grains are stuck together by ________ that grow from solution.

2 Clay cannot dry out under the sea. How is the water removed when clay is turned to shale?

3 Draw close-up diagrams to show how sand turns to sandstone.

4 In what way are sedimentary rocks recycled rocks?

5 Shale containing fossil sea shells is sometimes found high up in the mountains. Where did this sedimentary rock form?

Key ideas

Sedimentary rocks are made from the broken fragments of other rocks or the remains of living creatures.

The remains of living creatures found in sedimentary rocks are called **fossils.**

11e All change

Sedimentary rocks do not stay beneath the sea forever. Earth movements can lift them up to form cliffs, or even fold them into new mountains.

The same enormous forces can also push them deeper down into the crust of the Earth. There they get hotter and hotter and the pressure becomes more and more intense. Under these conditions, they start to change.

Fossils found at the tops of these mountains show that the rocks formed under the sea!

Bake it

Clay can easily be moulded into blocks which go hard when they dry. These blocks are fine for building in dry countries. But they wouldn't last long in Britain. A heavy rainstorm would soon turn them into mud.

If the clay blocks are baked in a hot oven called a kiln, they change into a completely new, weather-resistant material called brick.

A similar change happens to clay or shale if it is heated deep in the Earth's crust. A new, hard rock forms. Rock formed in this way is called **metamorphic rock**.

Squeeze it

Pressure changes clay to shale. But even greater pressure can change shale to **slate**. This is a very hard rock, just the thing to roof a brick house! The bricks of this house are more than 100 years old, but the natural slate on the roof is much older.

Baking and squeezing

Deeper in the Earth's crust, rocks are changed by both heat and pressure. This changes limestone to **marble**. Clay forms a metamorphic rock called **schist**. This sometimes contains semiprecious minerals like garnet.

schist

Around and around

As the rocks get hotter, the metamorphic changes become greater. If the rocks gets too hot they will melt. But if molten rock cools and sets, it becomes igneous rock.

So igneous rocks weather and are eroded to form sediments. Sediments harden to form sedimentary rock. Sedimentary rock can change to metamorphic rocks. And metamorphic rocks can melt and so turn back to igneous rock.

The rocks of the Earth can be recycled. This is known as the **rock cycle.**

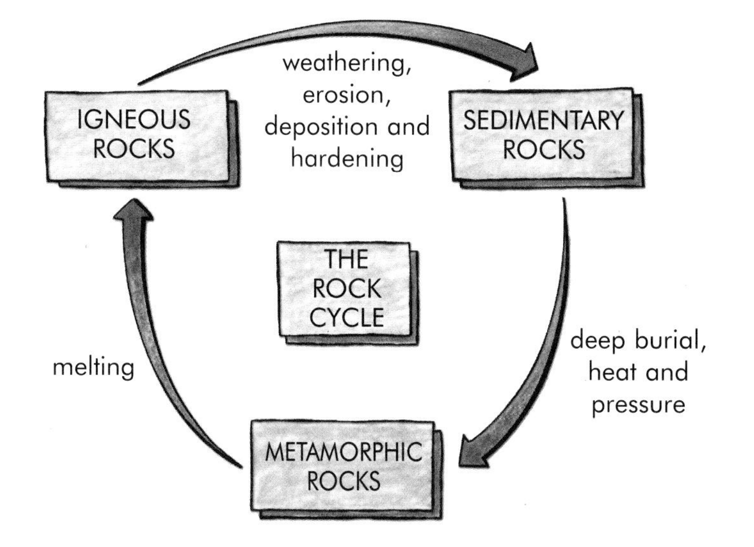

1 Copy the diagram of the rock cycle.

2 Add examples of igneous, sedimentary and metamorphic rocks to it.

3 Write an account of the travels of a grain of clay over 200 million years.

What do you know?

1 Copy and complete the following sentences. Use the words below to fill the gaps.

metamorphic recycled igneous sediments

The rocks of the Earth are naturally _______ in the rock cycle. Igneous rocks break down on the surface of the Earth, forming ________. Sedimentary rocks are changed into ________ rocks by heat and pressure. These rocks melt if they get too hot, turning back to ________ rock.

2 Bricks and slate are both made from clay. How was each formed?

3 Fossil sea shells are sometimes found at the tops of mountains, and schist is found at the surface of the Earth. What do these facts tell you about the Earth's crust?

Key ideas

Sedimentary rocks can be changed by heat and pressure into **metamorphic rocks**.

Metamorphic rocks can melt, cool and set to form igneous rocks, which can break down to form sedimentary rocks.

These changes make up the **rock cycle**.

11 EXTRAS

11a Exploding volcanoes

Some lava is full of bubbles, like an Aero chocolate bar. This is because the molten lava has gas dissolved in it, which bubbles out as the lava reaches the surface.

A similar thing happens when you open a can of fizzy drink.

Mount Etna has an open crater, so this gas can escape easily. But other volcanoes such as Mount Vesuvius in Italy are 'self-sealing'. Their lava freezes over between eruptions, plugging the crater and trapping the gas. This means that the gas builds up until the top of the volcano is blown off in a gigantic explosion.

This girl was killed by the eruption of Vesuvius. Her body was preserved in the volcanic ash for 2000 years.

During Roman times, a great city called Pompeii grew up close to Mount Vesuvius. Vesuvius had not erupted for centuries, so people thought it was completely inactive (extinct). In 79 AD, a great dark cloud suddenly appeared above the volcano. The earth shook and people heard an enormous explosion. As they ran out to see what was happening it began to rain droplets of molten rock, ash and dust. The entire population was killed almost at once and was buried by a thick layer of volcanic ash. The town lay undisturbed until it was discovered by archaeologists in the nineteenth century.

1 Explain what happened to Vesuvius in 79 AD.

2 Vesuvius has erupted every 50 years or so since 79 AD. Do you think its eruptions have been as powerful as the one that destroyed Pompeii? Explain your answer.

3 Vesuvius last erupted in 1945. Should the people living nearby be worried?

What happened to the river?

Life in the desert country of Klatch was concentrated on the fertile flood plains of its great river. In good years, the farms provided more than enough food to feed the people. But the river was a mixed blessing. Every year when heavy rain fell on the mountains, the river burst its banks and flooded the fields. At other times, drought reduced the river to a string of stagnant pools, and drinking water was scarce.

To overcome this problem, the river was dammed up. As a great reservoir lake filled up, the flow of water was controlled. There was now plenty of fresh water to drink all year. In the good years that followed, towns and cities spread along the valley.

Thirty years later, all is not so well. The farmers are complaining that their crops do not grow so well. On the delta, high tides and storms have washed away some coastal villages and salt sea water has killed the crops. The reservoir is beginning to fill up with clay and sand. The extra weight could burst the dam. What has gone wrong?

1 Why do the crops grow less well than they did before the dam was built?

2 Why is the lake now filling up with sand and clay?

3 Why is the sea beginning to wash over the delta?

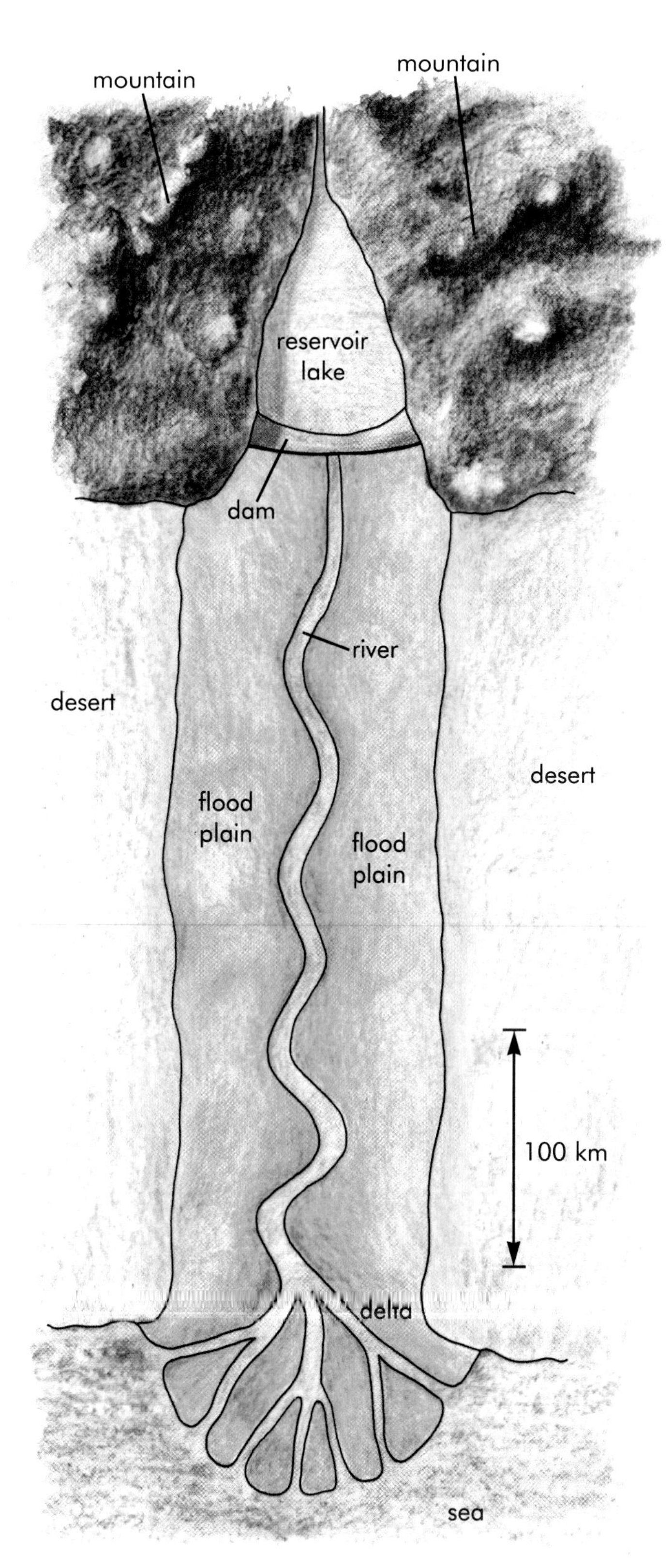

11d A slice through time

Sediments form layers on the sea bed, and these harden to sedimentary rock. The youngest layer is at the top. If you dig down through the layers, the rocks get older and older.

A fossil trilobite

Sometimes great Earth movements force these beds of sedimentary rock up out of the sea. Rivers then cut down through them, exposing the older beds as they go.

In Arizona, USA, the Colorado River has cut its way down through an enormous thickness of sedimentary rocks, forming the Grand Canyon.

Near the top of the canyon the rocks are 100 million years old. They contain the fossil traces of dinosaurs. Near the river they are 600 million years old and contain fossil trilobites. These strange sea creatures were amongst the first living things on Earth.

1 The rocks near the top of the canyon are softer than those near the bottom. Why?

2 Where was the top layer of rock formed?

3 What evidence is there that the oldest rocks in the canyon formed in the sea?

4 The sequence of rocks in the canyon took 500 million years to form. What is the longest possible time that the river has taken to cut through them?

11e Is it igneous or metamorphic?

Igneous and metamorphic rocks are both made from crystals. They may even be made from the same type of crystals. So how can you tell them apart? It's all in the way the crystals are arranged.

Igneous rocks form as a liquid rock sets. Needle-shaped or plate-like crystals can grow in any direction. Igneous rocks therefore have crystals scattered at random.

Most metamorphic rocks form when new crystals start to grow in solid rock under pressure. Needle-shaped or plate-like crystals are only free to grow at right angles to this pressure. This makes the rocks banded or layered.

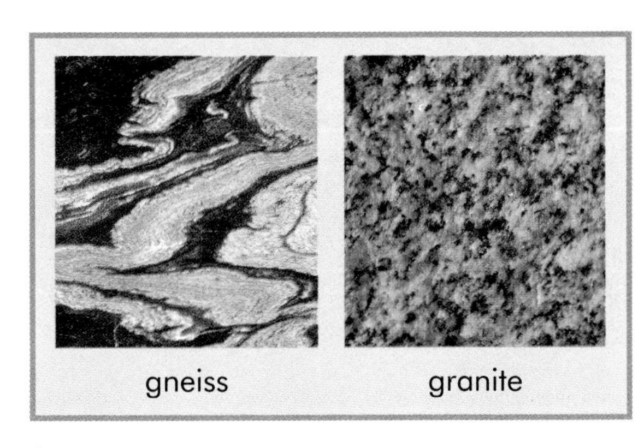

gneiss granite

1 Look at the two rocks shown. Which one is igneous and which is metamorphic? Explain your answer.

2 Lava can bake any sediment it flows over. Metamorphic rock formed like this does not show banding. Why not?

Glossary

air resistance pages **8–9** — the friction from air when you move through it.

ammeter pages **132–3** — instrument for measuring the electric current flowing round an electrical circuit.

ampere page **132** — electric current is measured in amperes.

boiling point pages **95, 96, 97** — the temperature at which a liquid boils.

carbohydrates pages **56, 60** — a type of food made by plants in their leaves.

carbon dioxide pages **56, 100, 102** — a gas produced during burning and respiration and used by plants to make food.

carnivores page **25** — animals that eat meat.

cells pages **18–19, 30, 106** — the 'building blocks' that all things are made of.

chemical changes pages **100–3, 105** — changes to a substance that are not reversible, e.g. a candle burning.

chemical energy pages **47, 48, 50, 52** — a type of stored energy, e.g. in batteries.

chloroplasts page **56** — packets of green colour in plant leaves which trap light energy to make food.

chromatography pages **78–9** — a way of finding out what's in a mixture of coloured substances.

circuit diagram pages **126, 127** — a map of an electrical circuit using special symbols for all the components.

classifying pages **20–9, 31** — putting all living things into groups.

component page **127** — a part of something, e.g. a light bulb is a component in an electrical circuit.

condensation pages **76–7, 104** — water vapour turns back into liquid water when it cools down.

conductor — *see* **electrical conductor** *and* **heat conductor**.

consumers page **65** — consumers (animals) eat other organisms because they cannot make their own food.

contraction pages **98–9, 140** — a material getting slightly smaller, often when it cools.

decanting page **71** — carefully pouring off one liquid from another liquid or a solid.

density pages **38, 39** — the mass of a 1 cm cube of something.

diet pages **60–3, 68** — the usual food and drink of a person or animal.

dispersal pages **110–11, 122** — the way fruits carry seeds away from their parent plant.

dissolve pages **72–3** — some solids dissolve in some liquids. They disappear but you can get them back later.

distillation page **77** — getting pure water from a solution by evaporating or boiling it and then condensing the steam back into liquid water.

dynamo page **51** — a machine that generates electricity when it is turned.

ejaculate page **117** — semen is pumped out of the penis.

elastic energy page **52** — energy stored up in something springy, e.g. rubber, springs, or a trampoline.

electric current pages **124–37** — this carries electrical energy around a circuit.

electrical circuit page **124** — electric current flowing around a circuit of wires and components.

electrical conductors pages **36, 125, 136** — materials through which electric current can pass.

electrical energy pages **50, 134–5** — energy carried by electric current.

electrical insulators pages **36, 125** — materials through which electric current cannot pass.

electrical resistance pages **130–1** — this makes it harder for current to flow round a circuit.

embryo page **109** — the tiny plant or animal that develops from a fertilised egg.

energy pages **44, 55, 101, 102** — this has many different forms, e.g. heat, movement, light and sound.

energy transfer pages **44, 46, 47, 50** — passing energy from one thing to another, or changing energy from one type to another, e.g. electrical energy changing into light energy.

energy transfer diagram pages **46, 47, 50** — a way of showing with arrows and labels how energy moves from one thing to another.

energy value page **48** — the amount of energy something has, particularly food.

erosion page **142** — the wearing or washing away of rocks by water or ice.

evaporation pages **74–5** — water turning to water vapour without boiling.

excretion pages **15, 63, 69** — getting rid of the waste produced by living organisms.

expansion pages **98–9, 105, 140** — a material getting slightly bigger, often when it is heated.

Fallopian tubes pages **117, 144** — part of a woman's reproductive organs.

fat page **60** — a type of food and a store of energy in animals and plants.

fertilisation pages **106, 109, 117** — a male sex cell joining with a female sex cell.

fetus page **120** — an animal developing in its mother's uterus.

fibre page **63** — part of our food which we can't digest and which keeps food moving through the gut.

filtrate page **71** — the clear liquid that you get when you filter something.

filtration page **71** — separating a solid from a liquid by filtering.

flowering plants pages **26, 27** — plants that reproduce using flowers.

food chain pages **65, 67, 69** — the links between different animals that feed on each other and on plants.

food web page **67** — a collection of interlinked food chains.

force pages **2–13** — a push or a pull, measured in newtons (N)

fossils
pages **145, 150**
shells or bones of dead organisms from millions of years ago that are found in sedimentary rocks.

friction
pages **6–9, 13**
a force which slows down movement.

fuel
pages **46–7, 48–9, 54, 102**
an energy store.

gas
pages **33, 39, 41, 94, 96**
a substance with no fixed shape or volume.

gravitational energy
page **53**
a store of energy which something has when it has been lifted up against the force of gravity.

gravity
pages **4–5, 53**
the force that pulls things down to the ground.

growth
pages **15, 68**
an animal or plant getting bigger.

habitat
page 66
the place where a plant or animal lives.

heat conductors
page **36**
materials which heat can pass through.

heat energy
pages **46, 54**
anything hot has heat energy.

heat insulators
page **36**
materials which heat cannot pass through.

herbivores
page 25
animals that eat plants.

hydrogen
page **102**
a gas which can be used as a fuel.

igneous rocks
pages **138–9, 147, 150**
rocks formed by melted rock cooling.

implanting
page **119**
a fertilised egg fixing itself to the lining of the uterus.

insoluble
page **73**
a solid that will not dissolve.

invertebrates
pages **20, 22–3, 31**
animals with no backbone, e.g. jellyfish, starfish, worms, molluscs and arthropods.

joule
page **135**
a measurement of energy.

joulemeter
page **135**
an instrument for measuring energy.

key
pages **28–9**
a way of putting animals and plants into the groups they belong to.

kilogram (kg)
page **5**
the mass of something is measured in kilograms.

kilojoule
pages **48, 49**
a thousand joules.

kinetic energy
page **45**
the energy of anything that is moving.

lever
pages **10–11**
an easy way of lifting heavy weights.

light energy
pages **47, 56–7**
light is a form of energy.

liquid
pages **33, 39, 41, 42, 71, 72–3, 94, 96**
a substance with a fixed volume but no fixed shape.

living organism
pages **14–31, 122**
a living thing that is a plant or an animal.

magnetic
page **37**
magnetic metals like iron and steel can be picked up by magnets.

mammals
pages **24, 25**
one of the groups of vertebrate animals.

mass
pages **5, 38**
the amount of matter in an object or an animal. Mass is measured in kilograms.

melting point pages **95, 96, 97, 138** — the temperature at which a solid gets hot enough to turn to a liquid.

menopause page **118** — time when a woman stops having her periods.

menstrual cycle page **118** — the monthly ripening of an egg and preparation of the uterus for pregnancy, which leads to the period.

metals page **36–7, 96, 105, 124, 125** — shiny materials which are good conductors of heat and electricity. Many metals are hard and strong and can be shaped in different ways, and some are magnetic.

metamorphic rock pages **146, 147, 150** — rock formed by other types of rock getting squashed and heated.

microscope page **18** — an instrument for seeing very small things.

minerals pages **60, 62–3** — substances found in rocks. Some are important for a healthy human diet, e.g. iron, potassium and calcium.

model pages **40–1** — a way of representing things to help you understand how they behave.

movement pages **15, 16** — all living things can move by themselves.

movement energy pages **44–5** — the energy of anything that is moving. Also called kinetic energy.

newton (N) pages **3, 4, 5, 12** — a measurement of force.

non-flowering plants page **26** — plants that do not produce flowers, e.g. conifers, ferns and mosses.

non-living pages **14, 15, 30** — things that do not have the seven properties of living organisms.

non-metals page **36** — materials that are good insulators of heat and electricity.

nutrition pages **15, 17, 60** — getting food and using it for growth, movement and energy.

offspring page **14** — young plants or animals produced by parent organisms.

organism pages **14–31, 122** — a living thing.

ovary pages **108, 110, 111, 114, 115, 118** — the place where the female sex cells (eggs) are made and kept.

ovulation page **118** — time when an egg leaves the ovary.

ovule page **106** — the name for the female sex cell in a plant.

ovum pages **115, 116, 117, 118, 119** — the name for the female sex cell in an animal.

oxygen pages **56, 102, 103** — a gas in the air that reacts with fuels when they burn, and which all living things use to respire.

parallel wiring page **129** — a circuit in which two components are wired side by side.

particles pages **40–1, 43** — tiny invisible pieces that everything is made up of.

penis pages **112, 113, 116** — the part of the male reproductive system that places sperm in the female's body.

period pages **115, 118** — the part of the menstrual cycle when the lining of the uterus is shed.

photosynthesis page **56** — the process plants use to make food.

physical changes pages **100–1** — changes to a substance that are reversible, e.g. candle wax melting.

placenta page **120** — the organ that provides food and oxygen for the fetus in the uterus and removes its waste products.

pollen pages **106, 107, 108** — the name for the male sex cells in plants.

pollination pages **108–9** — pollen landing on the stigma of a plant.

primates page **25** — one of the groups of mammals. Humans are primates.

producers page **65** — organisms (usually plants) that make their own food.

properties pages **32–3, 34–5, 40–1, 42** — what things are or what they can do.

protein page **60** — a type of food which helps the body grow and repair itself.

puberty pages **112–13, 114–15** — the stage of development when the sexual organs and the body become adult.

pubic hair pages **113, 115** — hair in the groin which men and women grow after puberty.

reflection pages **84, 90–1** — light or sound bouncing back from a surface.

reflectors pages **84, 91** — surfaces that bounce back light or sound.

reproduction pages **15, 106–23** — the process of animals or plants making more of themselves.

respiration page **15** — using oxygen to release energy from food.

reversible pages **95, 100** — a change that can change back.

rock cycle page **147** — the way rocks change from one sort to another. Some rocks go through all types and become the type of rock they were in the first place.

samples page **14** — small bits of something which we can study to tell us about the whole thing.

saturated page **73** — a solution that will dissolve no more solute is saturated.

sediment pages **71, 142–3** — the solid left in the bottom of a liquid because it is insoluble or because the solution is saturated.

sedimentary rock pages **144–7, 150** — rock formed when sand, mud or lime is squashed and hardened.

semen pages **112, 117** — sperm and the liquid they are carried in.

sensitivity page **15** — being aware of things happening, and responding to them.

series wiring page **128** — a circuit in which two components are wired end to end.

sex cells page **106** — the male and female cells involved in reproduction.

sex organs pages **112, 113, 114** — the organs of a plant or animal involved in reproduction.

sexual reproduction pages **106–23** — making new organisms using special cells from a male and a female.

solid pages **32, 34–5, 40, 42, 71, 72–3, 94, 96** — a substance with a fixed shape and a fixed volume.

soluble page **73** — if a solid dissolves in a liquid, it is soluble.

solute page **73** — the solid that dissolves in a liquid to make a solution.

solution pages **72–3, 74–5** — a solid dissolved in a liquid.

solvent pages **73, 80** — the liquid a solute dissolves in.

sound energy page **50** — energy in the form of sound.

sperm pages **106, 107, 112, 116, 117, 119** — the name for the male sex cells in an animal.

stamen page **108** — the place where pollen is made in a flowering plant.

stigma page **108** — the part of the plant's female reproductive organs where the pollen lands.

stored energy pages **47, 48, 50, 52, 53** — energy that is stored, e.g. in fuel, food and batteries.

streamlined page **8** — shaped to make the effect of friction less.

temperature pages **94–9, 138–9, 140** — how hot or cold something is.

testes pages **112, 113** — the place where sperm are made in male animals.

thermometer pages **94, 99** — an instrument for measuring temperature.

transferring energy — *see* **energy transfer**.

umbilical cord page **120** — the cord that attaches the baby to the placenta in the womb.

uterus pages **114, 115, 118, 119** — the place where a young baby grows in a female mammal (the womb).

variable resistor page **130** — a component that can change the amount of resistance in an electrical circuit, to change the current flowing.

vertebrates pages **20, 21, 24–5** — animals with backbones, e.g. fish, amphibians, birds, reptiles and mammals.

vibrate pages **82–3, 92** — make regular shaking movements, which are sometimes too small to see. Sounds are made by vibrations.

vitamins pages **60, 62–3** — substances needed in small amounts to keep animals (including people) healthy.

volume pages **32, 33, 38** — the space taken up by a thing or person.

water cycle page **81** — water from the sea evaporates into the air, cools and condenses back to water in clouds. The water then falls as rain which runs back to the sea, and so on.

water vapour pages **74, 76, 77** — water as a gas.

weathering pages **140–1** — breaking rocks down by rain, ice, heat, or chemical reactions.

weight pages **4–5, 34, 38–9, 43** — the pulling force of the Earth's gravity on something, measured in newtons.